中国世界级非遗文化悦读系列·寻语识遗
丛书主编 魏向清 刘润泽

中国传统桑蚕丝织技艺·缂丝

（汉英对照）

郭启新　赵传银　主编

Traditional Chinese Sericulture and Silk Craftsmanship: Kesi

南京大学出版社

本书为以下项目的部分成果：

南京大学外国语学院"双一流"学科建设项目

全国科学技术名词审定委员会重点项目"中国世界级非物质文化遗产术语英译及其译名规范化建设研究"

教育部学位中心 2022 年主题案例项目"术语识遗：基于术语多模态翻译的中国非物质文化遗产对外译介与国际传播"

南京大学-江苏省人民政府外事办公室对外话语创新研究基地项目

江苏省社科基金青年项目"江苏世界级非物质文化遗产术语翻译现状与优化策略研究"（19YYC008）

江苏省社科基金青年项目"江苏世界级非遗多模态双语术语库构建研究"（23YYC008）

南京大学暑期社会实践校级特别项目"讲好中国非遗故事"校园文化活动

参与人员名单

丛书主编 魏向清 刘润泽
主　　编 郭启新 赵传银
翻　　译 张学梓
译　　校 Zhujun Shu　Benjamin Zwolinski
学术顾问 许雅香
出版顾问 何宁　高方
中文审读专家（按姓氏拼音首字母排序）
　　　　　　陈俐　丁芳芳　王笑施
英文审读专家 Colin Mackerras　Leong Liew
参编人员（按姓氏拼音首字母排序）
　　　　　梁鹏程　孙文龙　王朝政　王曼琪　吴小芳
　　　　　叶莹　张璐謦　张亚萍　章玉兰
手　　绘 裴梓含
摄　　影 刘韶方
知识图谱 王朝政
中国历史纪年简表 王朝政
特别鸣谢 江苏省非物质文化遗产保护研究所
　　　　　范玉明缂丝大师工作室

编者前言

2019年秋天开启的这次"寻语识遗"之旅，我们师生同行，一路接力，终于抵达了第一个目的地。光阴荏苒，我们的初心、探索与坚持成为这5年奔忙的旅途中很特别，也很美好的回忆。回望这次旅程，所有的困难和克服困难的努力，如今都已经成为沿途最难忘的风景。

这期间，我们经历了前所未有的自主性文化传承的种种磨砺，创作与编译团队的坚韧与执着非同寻常。古人云，"唯其艰难，方显勇毅；唯其磨砺，始得玉成"。现在即将呈现给读者的是汉英双语对照版《中国世界级非遗文化悦读系列·寻语识遗》丛书（共10册）和中文版《中国世界级非遗文化悦读》（1册）。书中汇聚了江苏牵头申报的10项中国世界级非物质文化遗产项目内容，我们首次采用"术语"这一独特的认知线索，以对话体形式讲述中国非遗故事，更活泼生动地去诠释令我们无比自豪的中华非遗文化。

2003年，联合国教科文组织（UNESCO）第32届会议正式通过了《保护非物质文化遗产公约》（以下简称《公约》），人类非物质文化遗产保护与传承进入了全新的历史时期。20多年来，世界"文化多样性"和"人类创造力"得到前所未有的重视和保护。截至2023年12月,中国被列入《人类非物质文化遗产代表作名录》的项目数量位居世界之首(共43项),是名副其实的世界非遗大国。正如《公约》的主旨所述,非物质文化遗产是"文化多样性保护的熔炉,又是可持续发展的保证",中国非遗文化的世界分享与国际传播将为人类文化多样性注入强大的精神动力和丰富的实践内容。事实上,我国自古就重视非物质文化遗产的保护与传承。"收百世之阙文,采千载之遗韵",现今留存下来的卷帙浩繁的文化典籍便是记录和传承非物质文化遗产的重要载体。进入21世纪以来,中国政府以"昆曲"申遗为开端,拉开了非遗文化国际传播的大幕,中国非遗保护与传承进入国际化发展新阶段。各级政府部门、学界和业界等多方的积极努力得到了国际社会的高度认可,中国非遗文化正全面走向世界。然而,值得关注的是,虽然目前中国世界级非物质文化遗产的对外译介与国际传播实践非常活跃,但在译介理据与传播模式方面的创新意识有待加强,中国非遗文化的国际"传播力"仍有待进一步提升。

《中国世界级非遗文化悦读系列·寻语识遗》这套汉英双语丛书的编译就是我们为中国非遗文化走向世界所做的一次创新译介努力。该编译项目的缘起是南京大学翻译专业硕士教育中心特色课程"术语翻译"的教学实践与中国文化外译人才培养目标计划。我们秉持"以做促学"和"全过程培养"的教学理念,探索国别化高层次翻译专业人才培养的译者术语翻译能力提升模式,尝试走一条"教、学、研、产"相结合的翻译创新育人之路。从课堂的知识传授、学习，课后的合作研究，到翻译作品的最终产出，我们的教研创新探索结出了第一批果实。

汉英双语对照版丛书《中国世界级非遗文化悦读系列·寻语识遗》被列入江苏省"十四五"时期重点图书出版规划项目，这是对我们编译工作的莫大鼓励和鞭策。与此同时，我们受到来自国际中文教育领域多位专家顾问的启发与鼓励，又将丛书10册书的中文内容合并编成了一个合集《中国世界级非遗文化悦读》，旨在面向国际中文教育的广大师生。2023年夏天，我们这本合集的内容经教育部中外语言交流合作中心教研项目课堂试用，得到了非常积极的反馈。这使我们对将《中国世界级非遗文化悦读》用作非遗文化教材增添了信心。当然，这个中文合集版本也同样适用于国内青少年的非遗文化普及，能让他们在"悦读"过程中感受非遗文化的独特魅力。

汉英双语对照版丛书的编译理念是通过"术语"这一独特认知路径，以对话体形式编写术语故事脚本，带领读者去开启一个个"寻语识遗"的旅程。在每一段旅程中，读者可跟随故事里的主人公，循着非遗知识体系中核心术语的认知线索，去发现、去感受、去学习非遗的基本知识。这样的方式，既保留了非遗的本"真"知识，也彰显了非遗的至"善"取向，更能体现非遗的大"美"有形，是有助于深度理解中国非遗文化的一条新路。为了让读者更好地领会非遗知识之"真善美"，我们将通过二维码链接到"术语与翻译跨学科研究"公众号，计划陆续为所有的故事脚本提供汉语和英语朗读的音频，并附上由翻译硕士专业同学原创的英文短视频内容，逐步完成该丛书配套的多模态翻译传播内容。这其中更值得一提的是，我们已经为这套书配上了师生原创手绘的核心术语插图。这些非常独特的用心制作融入了当代中国青年对于中华优秀传统文化的理解与热爱。这些多模态呈现的内容与活泼的文字一起将术语承载的厚重知识内涵，以更加生动有趣的方式展现在读者面前，以更加"可爱"的方式讲好中国非遗故事。

早在10多年前，全国高校就响应北京大学发起的"非遗进校园"倡议，成立了各类非遗文化社团，并开展了很多有益的活动，初步提升了高校学生非遗文化学习的自觉意识。然而，我们发现，高校学生群体的非遗文化普及活动往往缺乏应有的知识深度，多限于一些浅层的体验性认知，远未达到文化自知的更高要求。我们所做的一项有关端午非遗文化的高校学生群体调研发现，大部分高校学生对于端午民俗的了解较为粗浅，相关非遗知识很是缺乏。试问，如果中国非遗文化不能"传下去"，又怎能"走出去"？而且，从根本上来说，没有对自身文化的充分认知，是谈不上文化自信的。"求木之长者，必固其根本；欲流之远者，必浚其泉源。"中国世界级非遗文化的对外译介与国际传播要解决的关键问题是培养国人尤其是青少年的非遗文化自知，形成真正意义上基于文化自知的文化自信，然后才有条件由内而外，加强非遗文化的对外译介与国际传播。非遗文化小书的创新编译过程正是南京大学"非遗进课堂"实践创新的成果，也是南大翻译学子学以致用、培养文化自信的过程。相信他们与老师一起探索与发现，创新与传承，译介与传播的"寻语识遗"之旅定会成为他们求学过程中一个重要的精神印记。

我们要感谢为这10个非遗项目提供专业支持的非遗研究与实践方面的专家，他们不仅给我们专业知识方面的指导和把关，而且也深深影响和激励着我们，一步一个脚印，探索出一条中国非遗文化"走出去"和"走进去"的译介之路。事实上，这次非常特别的"寻语识遗"之旅，正是因为有了越来越多的同行者而变得更加充满希望。最后，还要特别感谢南京大学外国语学院给了我们重要的出版支持，特别感谢所有参与其中的青年才俊，是他们的创意和智慧赋予了"寻语识遗"之旅始终向前的不竭动力。非遗文化悦读系列是一个开放的非遗译介实践成果系列，愿我们所开辟的这条"以译促知、以译传通"的中国非遗知识世界分享的实践之路上有越来越多的同路人，大家携手，一起为"全球文明倡议"的具体实施贡献更多的智慧与力量。

目 录
Contents

百字说明　A Brief Introduction

内容提要　Synopsis

知识图谱　Key Terms

桑基鱼塘　Mulberry-Fish Pond ·················· 001

采桑诗　Mulberry-Leaf-Picking Poems ·················· 010

桑梓之情　Love for Mulberries and Catalpas ·················· 025

蚕俗　Silkworm Customs ·················· 035

抱娘接　Mum-Hugging Grafting ·················· 049

采桑　Mulberry-Leaf Picking ·················· 058

催青　Artificial Incubation of Silkworm Eggs ·················· 065

炕床育　*Kang*-Bed Rearing ·················· 073

上蔟吐丝　Mounting and Spinning ·················· 084

煮茧缫丝　Cocoon Boiling and Silk Reeling ·················· 092

丝绸　Silk Fabrics ·················· 108

缂丝　Kesi ·················· 123

缂丝画和缂丝扇面　Kesi Paintings and Fans ·················· 132

通经断纬　Continuous Warp and Discontinuous Weft ……………141

结束语　Summary …………………………………………………155

中国历史纪年简表　A Brief Chronology of Chinese History ………157

百字说明

中国是世界上最早养蚕、制丝的国家,养蚕制丝至今已有5000多年的历史。中国传统桑蚕丝织技艺包括栽桑、养蚕、缫丝、织绸等工艺流程,是中国古代农耕文明与技术发展的重要体现。丝织品的流通开启了第一次东西方大规模的商贸往来通道,即"丝绸之路"。2009年,包括杭罗、宋锦和缂丝等在内的中国桑蚕丝织技艺被列入联合国教科文组织的《人类非物质文化遗产代表作名录》。

A Brief Introduction

The earliest practice of silkworm rearing and silk production in China dates back to over 5,000 years ago. The tradition of Chinese sericulture and silk craftsmanship incorporates the advances of agricultural technologies, which are applied in various procedures, including growing and caring for mulberry trees, raising silkworms, silk reeling and weaving silk threads, etc. Throughout history, the international trade of Chinese silk products facilitated the opening of the Silk Road, witnessing the first large-scale economic exchanges between the East and the West. In 2009, the major items of Chinese sericulture and silk craftsmanship, including Hangluo gauze, Song brocade, and Kesi, were inscribed on UNESCO's Representative List of the Intangible Cultural Heritage of Humanity.

内容提要

小龙的父亲龙教授是丝织研究专家。阳春三月，小龙邀请大卫随父亲一起去湖州游玩，并了解桑蚕文化习俗。接连成片的池塘和桑林引起了小龙和大卫的兴趣，他们与龙教授就此展开了有关桑蚕文化及技艺的讨论。接着，小龙和大卫参观了桑蚕养殖基地，实地了解采桑技术、养蚕流程以及缫丝过程。之后，小龙和大卫随龙教授到中国丝绸博物馆观看了缂丝展览。

Synopsis

Prof. Long, Xiaolong's father, is a silk-weaving researcher. In March, Xiaolong invited David to join them on a trip to Huzhou, a city near Hangzhou, to learn about sericulture customs. Curious about the continuous patches of ponds with mulberry trees around, Xiaolong and David consulted Prof. Long about sericulture and the craftsmanship involved. They gained a first-hand knowledge from their own experiences in a sericulture base about how to pick mulberry leaves, rear silkworms and extract silk. At last, they visited a Kesi exhibition at China National Silk Museum in Hangzhou.

知识图谱
Key Terms

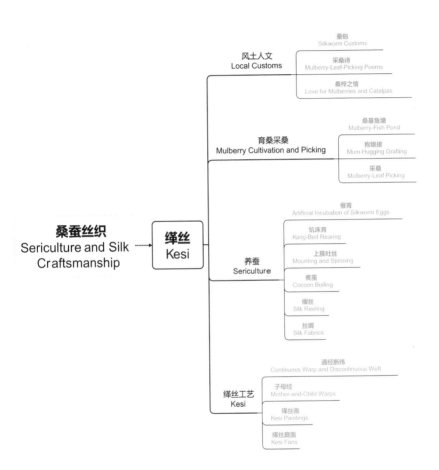

桑基鱼塘

阳春三月，小龙和大卫跟着龙教授来到湖州郊区游玩，并了解桑蚕文化习俗。

大　　卫：小龙，这里环境真不错，景色好美呀。

小　　龙：是啊，这里树很多，而且还有成片的池塘。

大　　卫：这些树都好矮，是什么树呀？

龙教授：是桑树，这是专门培养的矮化树种，为了方便采摘桑叶。几千年来，种桑养蚕一直是中国人重要的生产和生活方式。不过，现在桑树在城市里不常见了。

大　　卫：可能城市里没人养蚕吧。

龙教授：嗯。城里没有养蚕的资源和环境。你们看，住在农村，除了养蚕，还能养鱼。池塘边种桑树是这里的一种农业生产模式，叫"桑基鱼塘"。

桑基鱼塘
Mulberry-Fish Pond

小　龙：为什么叫桑基鱼塘？

龙教授：在地势低平、河网密布的地方开挖池塘养鱼，挖出来的泥土直接堆在鱼塘的四周作塘基，然后再在塘基上种桑树，这就是桑基鱼塘。

小　龙：这种一举两得的生产模式真是绝妙呀。

龙教授：桑基鱼塘是典型的中国式复合农业系统，用现在的时髦说法叫"良性循环的生态系统"。

大　卫：是怎么良性循环的呢？

龙教授：桑基鱼塘把蚕桑业和水产养殖业结合起来了。桑叶可以养蚕，蚕粪可以喂鱼，鱼粪可以肥塘，塘泥又可以肥桑。桑、蚕、鱼之间形成一种良性循环，构成可持续的农业经济和良好的生态系统。

小　龙：充分利用资源，这种理念很先进呀。

大　卫：龙教授，这种桑基鱼塘在南方很普遍吗？

龙教授：是的。经过上千年的发展，种桑养蚕和蓄水养鱼相辅相成，形成了非常特别的江南水乡农业生态。

小　龙：上千年？那就是说，桑基鱼塘这种模式很早就有了？

龙教授：是啊，有2500多年的历史呢。你们想象一下，当年湖州"处处倚蚕箔，家家下鱼筌"[①]的景象有多美。

大　卫：嗯，我喜欢乡村美景。

小　龙：这里面桑树很重要呀。有了桑树，就可以养蚕，蚕丝可以织漂亮的丝绸，难怪湖州的丝绸很出名呢。

龙教授：是啊，这小小的桑叶不可小瞧，它在古代可是重要的生产物资，还引发过战争呢。

大　　卫：引发战争?

龙教授：是的,《史记》②中就记录了这样一场战争。

小　　龙：爸爸,快给我们讲讲。

龙教授：2500多年前,吴国和楚国两国边境上的百姓为争采桑叶发生了纠纷,吴王知道后就派兵去攻打楚国。这是中国古代文献记载最早的"争桑之战"。

大　　卫：龙教授,采桑是不是有很多故事呀?

龙教授：有啊,还不少呢。

注释：

① "处处倚蚕箔,家家下鱼筌"：出自唐朝诗人陆龟蒙《奉和袭美太湖诗·崦里》。

②《史记》：西汉史学家司马迁(公元前145年?—公元前90年?)撰写的纪传体史书,是中国历史上第一部纪传体通史,记载了从上古传说中的黄帝时代到汉武帝太初四年间共3000多年的历史。

Mulberry-Fish Pond

It was a sunny day in March. David was invited to the outskirts of Huzhou with Xiaolong and Prof. Long, in the northern part of Zhejiang Province. They learnt about some sericulture customs.

David: Xiaolong, what a nice place! The scenery is wonderful!

Xiaolong: Yes. There're a lot of trees and a continuous expanse of ponds.

David: These trees are quite short. What are they?

Prof. Long: They're mulberry trees, specially-bred dwarf plants, friendly for picking mulberry leaves. For thousands of years, planting mulberry trees and rearing silkworms have been an important way of life for Chinese people. But today, you could barely see these plants in the city.

David: Maybe because no one raises silkworms in the city.

Prof. Long: You're right. You don't have the necessary resources and environment after all. But in rural areas, silkworm breeders can not only keep silkworms, but also raise fish. Did you see the ponds encircled by mulberry trees? They represent a local agricultural production model, known as the mulberry-fish pond system.

Xiaolong: How did it get the name?

Prof. Long: Across the flat terrain and within the dense river network, people dig ponds to raise fish and pile the excavated soil around the pond to form the dyke, on which mulberry trees are planted. So, the mulberry-fish pond is formed.

Xiaolong: What an ingenious idea! It can meet several ends.

Prof. Long: Yes. It's a typical compound agricultural system in China. To put it another way, it's a sustainable ecosystem.

David: How does it work?

Prof. Long: This model integrates sericulture with aquaculture.

To be more specific, mulberry leaves are used to raise silkworms, and silkworm feces are used to feed the fishes. Then, fish excrement nourishes the pond; and the pond mud serves as a natural fertilizer for mulberry plants. So, a virtuous cycle forms among mulberry trees, silkworms and fishes, constituting a sustainable agroecology characterised by a circular economy.

Xiaolong: The model makes the best of the local resources and shows an advanced agricultural conception.

David: Is the system common in southern China?

Prof. Long: Yes. Throughout the thousand-year development, the ecosystem in the Yangtze River Delta has been kept, where mulberry cultivation and sericulture interact with fish farming.

Xiaolong: Thousand-year development? The mulberry-fish pond model has existed for a very long time?

Prof. Long: Exactly. It has a history of more than 2,500 years. Just imagine, how beautiful Huzhou was in the verse "A line of silkworm baskets sit against walls;

A spread of bamboo fish traps worm their ways in rivers". [1]

David: Yes. The countryside landscape described in the verse is really beautiful. I love it.

Xiaolong: So do I. And the mulberry trees are not only beautiful, but also very useful. You know, with mulberry trees, silkworms can be raised and the fibres spun can be used to weave exquisite silk fabrics. No wonder Huzhou is famous for its silk.

Prof. Long: You're right. Mulberry trees aren't quite as ordinary as they look. In ancient times, they were literally significant production materials and even triggered a war.

David: A war?

Prof. Long: Well, it really happened, as it's recorded in *Shiji* [2].

Xiaolong: Dad, tell us more about it then.

Prof. Long: Well, over 2,500 years ago, a dispute broke out over picking mulberry leaves along the border between the states of Wu and Chu. The King of Wu sent troops to attack the State of Chu. This is the first written document about a mulberry war.

David: Wow. Are there any fascinating stories about mulberry-leaf picking?

Prof. Long: Of course. Quite a lot.

Notes:

1. The verse is a quotation from the poem "Responding to Ximei's Poems about Lake Tai" by the Tang poet Lu Guimeng.

2. *Shiji*: *The Records of the Grand Historian*, authored by Sima Qian (145B.C.?—90B.C.?), the great historian in the Western Han Dynasty. It is China's first biographical general history, documenting 3,000-odd years of history from the Yellow Emperor era in the ancient legend down to the fourth year of the Taichu period during the reign of Emperor Wu of the Han Dynasty.

采桑诗

中午的暖阳之下,小龙、大卫和龙教授三人坐在鱼塘边。他们一边休息,一边聊起了采桑的故事。

大　卫：龙教授,给我们说说采桑的故事吧。

龙教授："东风二月暖洋洋,江南处处蚕桑忙。"①

小　龙：爸爸,您怎么念起诗来了?

龙教授：古时候那些采桑的故事就记录在诗歌里呢。这是一首描写南方采桑情景的诗歌。你们听懂了吗?

小　龙：这里的"二月"是指中国农历的二月②吧。

大　卫：这是说天气暖和起来了,有东面吹来的风,大家忙着采桑养蚕,对吗?

龙教授：是的,这里东风特指春天的暖风,采桑就是从春天开始的。"蚕生春三月,春桑正含绿。女儿采春桑,歌吹当春曲。"③

小　　龙：这首诗描述春天蚕宝宝出来了，桑树发芽了，小姑娘唱着歌去采桑，对吗？

龙教授：没错。春天是开始采桑养蚕的时候。古代这样的采桑诗很多，比如这首《采桑曲》④：

青溪女儿爱罗裙，

提筐陌上踏春云。

蚕饥日暮思归去，

不敢回头看使君。

小　　龙：这首诗写得很有画面感。大卫，你能想象得出来吗？

大　　卫：这首诗我们中文课上讲过。我来讲一讲，你们看我理解得对不对？在温暖的春天里，穿着罗裙的年轻女孩提筐去采桑叶，黄昏采完桑叶赶紧回家喂蚕，都没时间回头看一眼她的男朋友，这样理解对吗？就是前面两个字"青溪"忘了是什么意思。

龙教授：大卫中文学得不错。"青溪"是水流清清的小溪。你们看，这首诗是不是很细致地描写了采桑女的生活？

小　　龙：是的。

大　　卫：我喜欢最后一句"不敢回头看使君"，说出了她心

采桑诗
Mulberry-Leaf-Picking Poems

罗敷喜蚕桑，
采桑城南隅。

里的秘密，很有趣。

龙教授：中国的蚕桑丝织历史悠久。种桑、养蚕、织布就是人们的日常生活，所以采桑成为中国古代诗歌的一大题材。采桑诗是蚕桑文化中极其重要的一部分。

采桑诗　Mulberry-Leaf-Picking Poems　013

采桑
Picking Mulberry Leaves

小　龙：爸爸，好像很多采桑诗都会提到男女之情，就像刚才那首一样。

大　卫：采桑诗也可以算作爱情诗吗？

龙教授：当然可以了。在古代，桑林也是男女约会的地方，所以很多采桑诗都会涉及男女之情。除此之外，采桑诗也真实再现了人们的生活状态。比如，这首《采桑》⑤是这样写的："朝去采桑日已曙，暮去采桑云欲雨。"

小　龙：这是描述蚕农们的采桑之苦吧？是说采桑要么顶着太阳，要么淋着雨。

龙教授：没错。我们是农业大国，耕织是古代社会的主要经济支柱，所以不光诗人写诗话桑麻，连皇帝也作采桑诗呢。

大　卫：皇帝又没在民间生活，他们怎么会写采桑诗呢？

龙教授：这正说明统治者对耕织极其重视呀。

小　龙：明白了，采桑纺织是国家大事，所以皇帝必须关心重视。

龙教授：是的。他们在诗词中提醒自己和官员要了解农事辛苦，鼓励百姓耕织。比如，清朝早期的康熙皇帝，就曾经写过采桑诗，这里面还有个故事呢。

大　卫：是皇帝亲自去采桑吗？

龙教授：那倒不是，是跟采桑有关的故事。康熙皇帝南巡时得到一本古人的《耕织图》图册。为了鼓励官员重视农事，他就让画家以此为范本，重新绘制了耕种图、纺织图各23幅，并在每幅画上附诗一首，还亲自为这本书写了序言，这就是著名的《御制耕织图》。通过这本书可以了解清朝农耕和丝织的情况。

小　龙：那就是关于耕织的绘本吧，是把整个农事的过程都画下来了吗？

龙教授：是的，耕种图把从浸种一直到收获、祭祀的整个过程都画了下来。

小　龙：那纺织图也应该是从种桑养蚕开始的吧？

龙教授：倒是没画种桑，是从采桑养蚕开始，最后是成衣。采桑图配的就是康熙皇帝的采桑诗。

大　卫：这样看来，采桑诗是很特别的一种诗歌了。

龙教授：是的，采桑养蚕是古人的日常生活，当然也寄托了他们的情感和思想。

大　卫：那我回去要找来读一读。

龙教授：好啊，读了采桑诗，才能更好地理解中国人的桑梓之情。

注释：

① "东风二月暖洋洋，江南处处蚕桑忙"：古典小说《醒世恒言》第十八卷"施润泽滩阙遇友"中的两句诗。

② 农历二月：根据中国传统历法，农历二月一般相当于阳历三月。

③ "蚕生春三月，春桑正含绿。女儿采春桑，歌吹当春曲"：出自南北朝民歌《采桑度》。

④《采桑曲》：明代沈天孙所作。

⑤《采桑》：宋代翁森所作。

Mulberry-Leaf-Picking Poems

In the warm sunlight at noon, Xiaolong, David and Prof. Long sat by a fish pond, talking about the stories of picking mulberry leaves.

David: Prof. Long, could you tell us some stories of picking mulberry leaves?

Prof. Long: "From east the warm February breeze blows; People along Southern Rivershore are active in the mulberry rows."[1]

Xiaolong: Dad, is that a poem?

Prof. Long: Yes. Many stories about picking mulberry leaves in ancient times were recorded in the form of poetry. This poem describes the people in southern China harvesting mulberry leaves. How do you understand this poem?

Xiaolong: The February² here follows the Chinese lunar calendar, right?

David: It describes the scene where the weather is warming up with the wind blowing from the east, and everyone is busy picking mulberry leaves, right?

Prof. Long: Yes. The east wind brings the warmth of spring here. The activities of gathering mulberry leaves usually begin in spring. Here's another poem. "Out in March silkworms are hatched, /When mulberry trees with green are matched. /Girls line out, merrily singing, /While gathering mulberry leaves, in praise of spring."³

Xiaolong: It describes a scene in spring, when baby silkworms come out of their eggs and mulberry trees sprout, girls start off picking mulberry leaves while singing songs on the way.

Prof. Long: Exactly. Spring is the beginning time for gathering mulberry leaves to feed silkworms. In fact, there are quite a lot of such poems. This "Song of Picking

Mulberry Leaves"[4] is another example: "A Qingxi girl, loving silk skirts to wear, /Baskets in hands, walks by spring clouds over there. /At dusk home she rushes, silkworms on mind; /No time to look back and her beloved one to find."

Xiaolong: The scene described in this poem is so vivid. David, can you imagine that?

David: Well, in fact I've learned about the poem in my Chinese class. According to my understanding, it is about a young girl in a silk skirt carrying a basket to pick mulberry leaves in a warm spring day. Then she hurried home to feed silkworms at dusk, and in such a hurry that she even didn't have time to turn back to see her boyfriend. Am I right? But I forgot what Qingxi means.

Prof. Long: Qingxi here refers to a stream with clear water. Isn't it a detailed description of her life?

Xiaolong: Yes.

David: I like the last line "No time to look back and her beloved one to find", because it reveals her secret thoughts. Interesting!

Prof. Long: China boasts a long history of sericulture and silk weaving. Growing mulberries, rearing silkworms and weaving fabrics were just part of ancient Chinese people's daily life. Therefore, mulberry-leaf picking has become a major topic of ancient Chinese poetry. It's fair to say that mulberry-leaf-picking poems count so much in China's culture of mulberry cultivation and silkworm raising.

Xiaolong: Dad, it seems that many such poems would touch upon the affection between men and women, like the above one.

David: Can they be counted as love poems?

Prof. Long: Of course. In ancient times, mulberry groves were often a dating place, so many mulberry-leaf-picking poems were related to the romance. In addition to this, such poetry also represented the real picture of people's daily life. For example, the poem "Picking Mulberry Leaves"[5] reads, "They leave to pick leaves at dawn, and then rises the sun. They leave to pick leaves at dusk, and then comes the rain."

Xiaolong: It tells the hardship of silkworm farmers, right? They have to work in both sunny and rainy days.

Prof. Long: You're right. Since China is a major agrarian country, farming and silk weaving used to be the main economic pillars. Even emperors in ancient China wrote poems on mulberry-leaf picking and farming.

David: But why? They didn't even live with the common people.

Prof. Long: It exactly shows the rulers valued farming and silk weaving.

Xiaolong: I see. These activities had a bearing on the national foundation and people's wellbeing, and thus emperors could not ignore them.

Prof. Long: Indeed. In their poems, they reminded themselves and officials to be aware of the hardships of farming and encouraged people to cultivate land and weave silk. For example, Emperor Kangxi of the early Qing Dynasty once wrote some mulberry-leaf-picking poems. There's also a story about it.

David: A story of Emperor Kangxi picking mulberry leaves himself?

Prof. Long: Of course not. It's a story about gathering mulberry leaves. In 1689, Emperor Kangxi, on an inspection tour in southern China, found a book *Pictures of Tilling and Weaving*. The book was handed down from previous dynasties. For the purpose of urging officials to take farming seriously, the emperor commissioned a painter to redraw 23 illustrations of tilling and another 23 illustrations of weaving based on the book, each inscribed with a poem and prefaced by the emperor himself. This is the famous *Pictures of Tilling and Weaving Made by Imperial Command*. It gives us a window into the rice cultivation and silk production during the Qing Dynasty.

Xiaolong: It's kind of a picture book of tilling and weaving? And does the tilling part cover every single activity related to rice production?

Prof. Long: Almost. It covers stages from seed soaking to

harvesting and making sacrifices.

Xiaolong: Then I guess the weaving part starts from mulberry cultivation and silkworm rearing?

Prof. Long: No, mulberry plantation is actually not included. It begins with mulberry-leaf picking and silkworm rearing, and ends with clothes making. Emperor Kangxi's poems on mulberry-leaf picking are attached to the pictures of the harvest scene.

David: It seems that the mulberry-leaf-picking poem is a special kind of poetry.

Prof. Long: Right. Gathering mulberry leaves and rearing silkworms was a part of the daily life for ancient Chinese people. It's no wonder that these poems carried their emotions and thoughts.

David: I'll find some to read later.

Prof. Long: Good. By reading those poems, we can better understand Chinese people's love for mulberries and catalpas.

Notes:

1. They are two lines from the eighteenth chapter "Shifu encounters a friend at Tanque" in China's classic fiction anthology *Stories to Awaken the World*.

2. Lunar calendar is the traditional Chinese time system. February on the lunar calendar is roughly equivalent to March on the solar calendar.

3. The verses are from a folk song in the Southern and Northern dynasties titled "Picking Mulberry Leaves".

4. **"Song of Picking Mulberry Leaves"**: It is written by the Ming writer Shen Tiansun.

5. **"Picking Mulberry Leaves"**: It is composed by the Song poet Weng Sen.

桑梓之情

> 大卫、小龙和龙教授三人继续坐在鱼塘边聊天。

大　卫：龙教授，刚才您说的"桑梓之情"是什么意思呢？

龙教授：在古代，人们普遍在房前屋后栽种桑树和梓树。

小　龙：那就是说，提到桑梓，就容易让人联想起家和亲人，是吧？

龙教授：没错，桑树和梓树与中国古代人的家庭生活息息相关。你们看，桑树的叶子可以养蚕，果实可以食用和酿酒。梓树生长速度快，很适合做家具。梓木历史上曾是刻板印书的好材料，因此我们也把稿件交付刊印称作"付梓"。

大　卫：原来在中国古代，桑树、梓树对一个家庭这么重要啊。

龙教授：是的，桑梓文化在中国历史悠久。桑梓常用来指代故乡。你们听这两句诗，"乡禽何事亦来此，令我生心忆桑梓"[①]。

小　龙：这是说作者漂泊在外，看到熟悉的禽鸟，想起故乡，想起家，是吧？

龙教授：是啊。人在他乡，特别容易思念故乡。桑梓也可用

桑梓之情　Love for Mulberries and Catalpas

桑梓之地　父母之邦　Where There Are Mulberries and Catalpas, There Is Hometown.

来指代父母。它还有另一层含义，就是象征生命再生。这两种树生长速度快、生命力极强，也被种植在墓地周围，作为生命再生的象征。

大　卫：这两种树的文化含义真丰富啊。

龙教授：古时候，桑梓在人们的生活中不可缺少，很自然会用桑梓指代故乡。

大　卫：我明白了。

龙教授：后来，"桑梓"一词又演变成故乡人或同族人的代称。这说明桑梓之情植根于中国人的内心深处。

小　　龙：真没想到桑树对中国文化的影响这么深远。除了表达对父母、家乡和乡亲的眷恋，桑树还有别的寓意吗？

龙教授：它还表达一种理想的田园生活状态。比如，像"把酒话桑麻"②和"鸡鸣桑树颠"③这样的诗句，反映了古人对自由、平静的田园生活的向往。

大　　卫：这种生活听起来真美好啊。

小　　龙：那桑树对我们现代生活还有影响吗？

龙教授：有啊。虽然已经21世纪了，织造技术日新月异，发展迅速，但人们种桑、采桑、养蚕、织丝绸的传统还在延续。中国许多地方，尤其是气候湿润的南方还是会大面积种桑。

小　　龙：我们找个时间去看看怎样采桑养蚕吧？

大　　卫：我特别想看看那些蚕宝宝是怎么吐丝的。

龙教授：好，今天下午我先带你们去了解一下湖州蚕俗，以后找时间再去桑蚕基地参观。

注释：

① "乡禽何事亦来此，令我生心忆桑梓"：出自唐朝诗人柳宗元（773—819）的《闻黄鹂》。

② "把酒话桑麻"：出自唐朝诗人孟浩然（689—740）的《过故人庄》。

③ "鸡鸣桑树颠"：出自东晋诗人陶渊明（365—427）的《归园田居·其一》。

Love for Mulberries and Catalpas

> David, Xiaolong and Prof. Long continued to chat by the mulberry-fish pond.

Daivd: Prof. Long, you mentioned the love for mulberries and catalpas just now. Does the phrase have any special meanings?

Prof. Long: Mulberry and catalpa trees were commonly planted around the houses in ancient times.

Xiaolong: Does it mean people tend to associate mulberries and catalpas with their homes and families?

Prof. Long: Yes. These two types of trees are strongly linked to the family life of ancient Chinese people. You see, mulberry leaves can be used to raise silkworms; the fruits can be eaten or made into wine. Catalpas grow fast and are suitable for making furniture. Also catalpas were good for woodblock engraving and printing books, so we still say *fuzi*（付梓）, which means literally "putting on to the catalpa block" when a manuscript goes to press.

David: Well, I didn't really know that mulberries and catalpas had been very important for a family in ancient times.

Prof. Long: That's true. And it has become a part of Chinese culture since long ago. Mulberries and catalpas are usually synonymous to native places, as can be seen from the two lines, "Orioles, why do you come here? I can't help but think of the mulberries and catalpas at home."[1]

Xiaolong: In the poem, the poet away from home recalled his family and hometown upon the sight of familiar birds.

Prof. Long: Indeed. Out of one's hometown, it's easy to be homesick. Mulberries and catalpas are also associated with parents. Another meaning is the regeneration of life. Fast-growing and strongly vigorous, they are also planted around tombs as a symbol of life regeneration.

David: The cultural connotations of the two kinds of trees are really rich and profound.

Prof. Long: Exactly. Mulberries and catalpas are indispensable for ancient Chinese people. So, they gradually become a symbol for people's hometown.

David: Well, I really got it.

Prof. Long: Later on, the phrase "mulberries and catalpas" is often used to denote fellow countrymen or kinsmen. It shows that mulberries and catalpas are cherished deeply by Chinese people.

Xiaolong: It never occurred to me that mulberry trees have such a far-reaching impact on Chinese culture. Apart from expressing people's attachment to their parents, hometowns and fellow countrymen, do they carry something else?

Prof. Long: Yes. They're also used to symbolise an ideal rural life. For example, we have such lines as "O'er wine we talk of mulberries and farming"[2] and "A rooster cries out atop a mulberry"[3], which present a free and tranquil country life that the ancients aspired to.

David: That kind of life sounds wonderful.

Xiaolong: But do mulberry trees still influence our modern life?

Prof. Long: Certainly. Although the 21st century has witnessed the rapid development of weaving technology, the tradition of mulberry cultivation, mulberry leaves gathering, silkworm rearing, and silk weaving is still being carried on. Extensive mulberry fields still exist in many parts of China, especially in the south where the climate is humid.

Xiaolong: Let's find some time to see mulberry leaves gathering and silkworm rearing?

David: I want to see how the silkworms start spinning.

Prof. Long: OK. This afternoon I'll take you to Huzhou to learn about some silkworm customs. And we'll plan the trip to a sericulture base for another time.

Notes:

1. The lines are from "Hearing the Calls of Orioles" by the Tang poet Liu Zongyuan (773–819).

2. It is from "Visiting an Old Friend's Cottage" by the Tang poet Meng Haoran (689–740).

3. It is from "Returning to Dwell in Gardens and Fields" by the poet Tao Yuanming (365–427) of the Eastern Jin Dynasty.

蚕俗

> 下午，大卫、小龙和龙教授一起来到湖州的衣裳街历史文化街区了解蚕俗。

小　　龙：爸爸，衣裳街这个名字好特别呀。

大　　卫：龙教授，这个街道名字有什么故事吗？

龙教授：有的，湖州是名副其实的丝绸之乡，这里发掘过4500年前的丝绢残片，是我国发现年代最早的丝织品实物，也是当今世界上最早的人工织物。

大　　卫：哇，那可是很久很久以前的东西。

龙教授：是的。这个地方的建城史有2000多年了。这里的丝绸贸易历史也很悠久。清朝指定湖州丝绸为宫廷专用衣料。200年前，这里就是生丝和绸缎的贸易集散地。英国、美国、日本等国的商人都来这里采购生丝，然后经宁波、广州转运回国。

小　　龙：难怪会有"衣裳街"这样的街名了。

龙教授：这里的许多地名都与桑蚕丝织有关，路上你们注意到没有？

小　　龙：我刚才看到了"育桑桥"。

龙教授：还有"迎锦桥"呢。这里有个地方叫"织里"，有条河叫"墨浪河"，这些都和丝织业有关。

大　　卫：墨浪河不是应该和墨水有关吗？怎么和丝织有关呢？

龙教授：之所以叫墨浪河，是因为当时印染丝绸的水把河水变成了墨黑色。

蚕俗
Silkworm Customs

小　　龙：爸爸，湖州这个地方为什么种桑、养蚕的特别多呢？

龙教授：湖州这里气候湿润，特别适合种桑养蚕。这里出产优质丝绸，素有"湖桑遍天下、湖丝甲天下"的说法。1851年，湖州的生丝在伦敦世界博览会上还拿过金奖呢。

大　　卫：噢，怪不得您带我们到湖州了解蚕桑文化习俗呢。

龙教授：是的，这里有独特的湖州蚕俗，还有"蚕神"的传说。

大　　卫：什么是蚕神？

龙教授：民间信奉各种各样的神，不同的神管的事情不一样，蚕神管的是和养蚕有关的事务。传说中教人种桑养蚕的嫘祖①就是蚕神之一。湖州当地流传最广的是含山的"蚕花公主"传说。

小　　龙：爸爸，给我们讲讲蚕花公主的故事吧。

龙教授：好啊。传说蚕花公主就住在含山脚下，因父亲外出打仗被困，蚕花公主就许愿说："谁能救我父亲，我就嫁给谁。"此时一匹白马飞奔过来，前去救回了她父亲。蚕花公主心存感激，准备嫁给白马，可是蚕

蚕花公主
Princess Silkworm

花公主的父亲觉得一匹马怎么可能配得上公主,便将蚕花公主的侍女嫁给了白马。

大　卫:怎么这么不讲信用呢?后来怎么样了?

龙教授:后来,马儿乱蹦撞死了侍女,于是家里人就杀了白马。马死后,公主很伤心,觉得对不起白马,就上吊自杀了。公主的坟上长出了一棵桑树,白马的坟上出现了很多蚕宝宝,蚕宝宝爬到桑树上吃桑叶。

因此民间就叫公主马头娘或者蚕花娘娘，还建了寺庙供奉公主和白马。久而久之，就形成了祭拜蚕花公主的习俗。

小　　龙：这个故事好凄惨呀。都怪这个父亲不讲信用！

龙教授：这个传说告诫人们要信守诺言。湖州还有其他蚕俗，比如蚕花生日、点蚕花火、轧蚕花、谢蚕花等。

小　　龙：我知道蚕花生日，好像是农历腊月十二，这一天蚕农们会备好酒菜祭祀蚕花娘娘。

大　　卫：那点蚕花火是点灯还是放焰火呢？

龙教授：点蚕花火的习俗是，除夕到大年初一早晨，蚕农在家里或庙里点油灯或蜡烛。

小　　龙：这像是蚕农版的年三十守夜了。那轧蚕花是什么呢？

龙教授：传统的轧蚕花是每年的清明②时期蚕农聚集在庙会上祭拜蚕神。"轧"是挤的意思。这一天的庙会人越多越拥挤越好。轧蚕花的活动以含山最有名。

大　　卫：为什么大家都要去那儿呢？

龙教授：传说清明当天，蚕花娘娘会化作村姑出现，留下蚕花喜气，蚕农都想把蚕花喜气带回家，让一年的种桑养蚕顺利，收获多多的蚕茧。

小　　龙：大家都想沾沾蚕花娘娘的喜气啊。

大　　卫：那谢蚕花就是感谢蚕花娘娘了，是吗？

龙教授：是的，蚕农用丰盛的酒菜表达对蚕花娘娘的感激，也有的在河边用泼水的形式进行祭祀。这些都是蚕农在感谢蚕花娘娘、庆祝蚕茧丰收。

小　　龙：真不愧是蚕乡，这么多的蚕俗。爸爸，下次带我们去桑蚕基地参观体验吧。

龙教授：行啊，过两天就带你们去。

注释：

① 嫘祖：传说中黄帝的妻子，发明了养蚕技术。

② 清明：二十四节气之一。每年阳历4月5日前后进入清明节气。祭祖、踏青是重要的清明节习俗。

Silkworm Customs

> In the afternoon, along with Xiaolong and Prof. Long, David came to the Yishang (衣裳, literally "clothes" in Chinese) Street, a historical and cultural district of Huzhou, to learn about the silkworm customs there.

Xiaolong: Dad, the name of Yishang Street is really special.

David: Prof. Long, is there any story about the name?

Prof. Long: Yes. Huzhou deserves the title of "the Home of Silk". Some silk fragments over 4,500 years ago were excavated here. They're the oldest fabric found in China and the world's earliest textile samples.

David: Wow. That's really a long time.

Prof. Long: Yes. Huzhou city itself has a history of more than

2,000 years. The silk trade in Huzhou also dates far back. During the Qing Dynasty, Huzhou silk fabrics were designated as imperial clothing materials. And 200 years ago, Huzhou was still a trading hub of raw silk and satin. Even the businessmen from other countries like the UK, the US and Japan, came here to purchase raw silk and then transported them home via the cities of Ningbo and Guangzhou.

Xiaolong: No wonder it's called the "Yishang Street".

Prof. Long: Names of many places here are related to silk weaving. Did you notice some on your way?

Xiaolong: I saw the "Yusang（育桑）Bridge". Yusang means mulberry cultivation.

Prof. Long: And the "Yingjin（迎锦）Bridge", or the Brocade Business Bridge. There's a weavers' community called "Zhili（织里）" and a river nearby called "Ink River", which are all related to silk weaving industry.

David: Isn't Ink River related to ink? How does it relate to silk manufacturing?

Prof. Long: It got its name because the water dyeing silk turned black.

Xiaolong: Dad, why does Huzhou have so many places for mulberry plantation and silkworm rearing?

Prof. Long: The humid climate here is suitable for planting mulberry trees and keeping silkworms. There has long been a local saying that "Huzhou mulberry travels the world; Huzhou silk leads the world". The high-quality raw silk from Huzhou won a gold medal at the London World Exposition in 1851.

David: Oh, no wonder you bring us here to learn more about the sericulture and related customs.

Prof. Long: Yes, Huzhou is known for its unique silkworm customs. And there's also a legend about the Silkworm Goddess.

David: Who is the Silkworm Goddess?

Prof. Long: According to the folklore, there're a variety of gods and goddesses who take charge of different aspects of people's life, and the Silkworm Goddess is in charge of silkworm rearing. Legend has it that Lady

Leizu[1] is the Silkworm Goddess who taught people to plant mulberry trees and rear silkworms. But the most widely recognised silkworm goddess in Huzhou is Princess Silkworm.

Xiaolong: Dad, tell us the story of Princess Silkworm, please.

Prof. Long: OK. According to the legend, Princess Silkworm lived at the foot of Hanshan. Her father was once trapped when fighting away from home. In desperation, the princess cried out to make a wish, "Whoever saves my father would be my husband." At that point, a white stallion galloped towards her and later brought her father back. The princess was grateful for the white horse and intended to marry him. But her father thought that the horse was way below his daughter. As a result, he married the princess's maid to the horse.

David: How could the father break the promise? What happened later?

Prof. Long: The stallion jumped frenetically and killed the maid. Consequently, the horse was killed by the family. Princess Silkworm hanged herself as she was heartbroken and felt sorry for the white horse. Then, surprisingly, a mulberry tree grew on her grave, and many baby silkworms appeared on the grave of the stallion. They climbed the tree to eat the leaves. Therefore, people named the princess the Horsehead Woman or Silkworm Goddess. And a temple was built in memory of Princess Silkworm and the white stallion. Over time, the local custom of worshipping Princess Silkworm came into being.

Xiaolong: What a tragic story! It's all on her father for breaking the promise.

Prof. Long: It also serves as a cautionary tale to emphasise the importance of keeping your words. And there're other silkworm customs in Huzhou, just to name a few, celebrating the birthday of Silkworm Goddess, holding up a light for Silkworm Goddess, worshiping Silkworm Goddess, rewarding Silkworm Goddess and so on.

Xiaolong: I know the first one. The birthday of Silkworm

Goddess seems to fall on the twelfth day of the twelfth month in the lunar calendar. Silkworm farmers will prepare wine and food as offerings for her.

David: Is the light in the second activity from a lamp or firework?

Prof. Long: Holding up a light for Silkworm Goddess means silkworm farmers keep oil lamps or candles at home or in the temple from New Year's Eve to the morning of the first day of the Lunar New Year.

Xiaolong: It's kind of silkworm breeders' version of staying up late on New Year's Eve. Then, what's the custom of worshiping Silkworm Goddess?

Prof. Long: Traditionally, silkworm farmers gather every year during Qingming Festival[2] at a temple fair to worship the Silkworm Goddess. It's believed that the more crowded the temple is, the better. The custom of worshipping Silkworm Goddess in Hanshan county is the most renowned one.

David: Why does everybody attend the temple fair?

Prof. Long: According to legend, on the very day of Qingming

Festival, Silkworm Goddess would appear as a village girl and give her blessings. The blessings would bring a successful year in silkworm raising and a good harvest of silkworm cocoons.

Xiaolong: It seems that everyone wants to be blessed by Silkworm Goddess.

David: Then what about the custom of rewarding Silkworm Goddess? Does it simply mean expressing thanks to Silkworm Goddess?

Prof. Long: Right. Silkworm farmers for this occasion prepare a substantial meal to express their gratitude to Silkworm Goddess. Some will worship the goddess by splashing water by rivers. The purposes of these activities are to thank Silkworm Goddess and to celebrate a good harvest.

Xiaolong: With so many silkworm customs, Huzhou is indeed the home of silkworms. Dad, please take us to a sericulture base next time.

Prof. Long: OK. We'll go there later.

Notes:

1. Lady Leizu: A legendary figure and the principal wife of the Yellow Emperor. It is said that Leizu invented silkworm rearing techniques.

2. Qingming Festival: Qingming is one of the 24 solar terms, which falls around April 5 on the Gregorian calendar. During Qingming Festival, people pay offerings to their ancestors and go out for spring sightseeing.

抱娘接

> 几天后,龙教授带着小龙和大卫来到桑蚕养殖基地参观桑树嫁接技术。

龙教授:桑树培育有多种方法,今天我们要看的是嫁接。嫁接就是把带芽的桑树枝条接到母树上,这样新枝很快就会长出桑叶来。1000多年前,湖州就已经广泛采用桑树嫁接技术了。

大 卫:龙教授,为什么要嫁接桑树呢?让它自己长大不就行了吗?

龙教授:嫁接可以促进桑叶的生长,也便于改良品种。这对桑叶的大规模生产非常重要。

小 龙:是不是嫁接后会很快长出更多的桑叶?

龙教授:是的。必须保证有足够的桑叶,才能养更多的蚕,结出更多的蚕茧,这样才可以实现蚕丝的批量

抱娘接　Mum-Hugging Grafting

生产。

大　卫：明白了。就是要给很多蚕宝宝足够的桑叶吃。

小　龙：那嫁接之后，桑树多长时间能长出新桑叶呢？

龙教授：半个月左右。

小　龙：这么快呀。爸爸，桑树随时都可以嫁接吗？有没有时间限制？

龙教授：嫁接一般是在3月中旬到4月上旬，也可以在夏天或秋天进行。我们来的时间正好。你们看，这就是刚嫁接过的桑树。

大　卫：咦，快看，这些树上都绑了塑料条，像是绑了绷带一样。

龙教授：是的，上面带芽的枝条叫"接穗"，下面这个母树叫"砧木"。把接穗和砧木接在一起，过一段时间这些接穗长大，桑树就会枝繁叶茂了。

大　卫：好神奇呀。简单地在树上切个口，接上新枝条就嫁接好了。

龙教授：桑树嫁接，看上去简单，实际并不简单，要讲究方法的。

小　龙：那怎么嫁接呢？

龙教授：嫁接桑树的方式有很多种，比如"袋接"和"抱娘接"。

大　卫："抱娘接"？这个名字有趣。

龙教授：抱娘接一般用于更换桑树品种或让老树更新复壮。我们来看看这位师傅的接法。

小　龙：把砧木切个倒U形切口，好像切得不深呀。

龙教授：是的，切断皮层就行了。如果砧木太粗，也可切A形切口。他手里拿的细芽就是接穗条。

大　卫：把它插进去就行了吗？

龙教授：是的。要选择健壮的带芽接穗条，插到切好的砧木切口中。

大　卫：他们操作得好熟练啊。

龙教授：当然了。要抓紧时间，不能误了农时。砧木根系发达，嫁接后复活率高，长势也旺。接穗芽变长后，要将砧木的上部锯断，保证新芽成长。

小　龙：怪不得这些树看上去很奇怪，很粗壮的树上枝条却很细，原来是嫁接的呀。

大　卫：龙教授，这个方法为什么叫"抱娘接"呢？

龙教授：你们仔细看看用作砧木的母树，还有上面的接穗。从背面看，这个切口像不像两条胳膊在抱着母树？

大　卫：很像。我明白了，叫"抱娘接"是因为这个切口的形状像孩子抱着妈妈的样子。这个叫法真形象呀。

Mum-Hugging Grafting

> Several days later, Prof. Long took Xiaolong and David to a sericulture base to learn more about mulberry grafting techniques.

Prof. Long: There're many ways to cultivate mulberry trees, and today we're going to see how to graft them. Grafting happens when mulberry branches with buds are joined onto the mother tree, and then the new shoots will soon grow mulberry leaves. The technology was widely used in Huzhou as early as about 1,000 years ago.

David: Prof. Long, what's the purpose of grafting? Why don't we let them grow up by themselves?

Prof. Long: Well, it can increase the harvest of mulberry leaves and allow for the improvement of varieties. It's important for the mass cultivation of mulberry leaves.

Xiaolong: Mulberry trees will soon grow more leaves after grafting?

Prof. Long: Sure. It's necessary to ensure an adequate supply of mulberry leaves, which is essential to rearing more silkworms, and producing more cocoons. In this way, the mass production of silk can be realised.

David: I see. We need enough mulberry leaves to feed silkworms.

Xiaolong: How long does it take for a mulberry tree to grow new leaves after grafting?

Prof. Long: Around half a month.

Xiaolong: So fast. Dad, can mulberry trees be grafted at any time? Or is there a specific time?

Prof. Long: Grafting is usually better carried out between mid-March and early April, and summer or autumn is also good for grafting. We come at the right time. Look, this is a freshly-grafted mulberry tree.

David: Let me see. These branches are bound with plastic strips, like bandages.

Prof. Long: Yes. The branches with buds are called the scion and the mother tree below is called the rootstock. After they join together the scion and the rootstock, the grafted mulberry tree will flourish after the scion piece grows long.

David: Amazing. They just make a cut on the tree and insert a new branch into it, then the grafting is done.

Prof. Long: It looks easy to graft mulberry trees, but in fact the process is much more complicated. It requires techniques.

Xiaolong: What are the specific techniques of grafting?

Prof. Long: Bark grafting, mum-hugging grafting, and so on.

David: Mum-hugging grafting? What an interesting name.

Prof. Long: It's usually adopted when there's a need to cultivate new varieties or to restore the vitality of old trees. Let's see how the worker performs the grafting.

Xiaolong: He makes an inverted U-shaped cut on the rootstock. It seems not a deep cut.

Prof. Long: Right. Just make sure to cut into the cortex. If the rootstock is too thick, an A-shaped cut is also fine. Look, the tender shoot in his hands is the scion piece.

David: Just insert it into the rootstock?

Prof. Long: Yes. People would select a robust and healthy scion with buds and then insert it into a downward cut in the rootstock.

David: People here are really adept at the technique of grafting.

Prof. Long: Of course. It's essential to act fast so as not to miss the growing season. The rootstock has a well-developed root system, and after the grafting there's a high probability for it to grow again and thrive. When the scion piece grows long, the upper part of the rootstock should be sawed off to make room for the growth of the grafted shoots.

Xiaolong: No wonder these trees look strange. The rootstock stem is much larger in size than the twigs on it. It turns out that they are grafted trees.

David: Prof. Long, why is the technique named mum-hugging grafting?

Prof. Long: Take a closer look at the mother tree used as the rootstock and the scion on it. And then see this cut from the back. Does it look like two arms around the mother tree?

David: Exactly. I get it. It's because the shape of the cut looks like a child hugging his or her mother. It's really a good name.

采桑

> 采桑的时节很快到了。一天早上,大卫、小龙和龙教授又来到桑蚕养殖基地。

小　　龙：爸爸,能不能带我们去采桑叶呀?

龙教授：可以呀。但桑叶可不是随便就摘的,要先了解方法。

大　　卫：采桑叶不是很简单的事吗?

小　　龙：对呀,这个不需要技术吧?

龙教授：当然不是这样,没有技术可不行。我先给你们普及一下养蚕常识吧。一千克鲜蚕茧大概有500个,那我问你们,出一千克蚕茧,得用多少桑叶?

小　　龙：这个可不好猜。也许10千克?

龙教授：还得再加上一半。

大　　卫：15千克。那还挺多呀。

龙教授：你们看，嫁接是为了多产桑叶，采桑叶是不是也要想办法维护桑叶的高产呀？

大　卫：是啊，那应该怎么做呢？

龙教授：既要考虑多采桑叶喂饱蚕宝宝，也要注意采摘桑叶对桑树生长的影响，所以采摘时，一定要有正确的方法。

小　龙：哦，就是不要一下子把桑叶都摘光，是吧？

龙教授：不错，不过不同季节采桑的方法也不一样。

大　卫：我原来以为只有春季采桑叶呢。

龙教授：当然不止春季。除了春叶，还有夏叶和秋叶。采桑方法根据养蚕的不同时期和桑树在各个季节的生长情况而定。

大　卫：都有哪些方法呢？

龙教授：基本方法有三种，就是摘叶法、采芽叶法和剪条法。

小　龙：摘叶法应该就是摘叶子，采芽叶法是只摘新发的桑叶，是吗？

龙教授：你说的第一种方法是对的，但第二种就不对了，这里说的芽叶是叶片和新梢的总称。

小　龙：采芽叶法是不是在春季用的比较多呀？

龙教授：是的。

大　卫：那剪条法呢？

龙教授：剪条法是把桑叶和枝条一起剪下来，春夏之交用的多。

大　卫：看来采桑叶也不简单呢。

龙教授：没错。错误的采桑方法会严重影响桑叶的产量。采桑的时间也是有讲究的，一般早晚两次，早上最好的时间段是6点到9点，下午在4点到6点之间。种桑养蚕得知道不同时间该做什么事，该用什么方法。等下先请采桑师傅教你们，然后你们学着采桑。

大　卫：我得好好学习。

小　龙：那我们赶紧去吧，已经8点了。

龙教授：好，我们先去体验一下春季采桑吧。

Mulberry-Leaf Picking

> Soon it's the time to pick mulberry leaves. One morning, David, Xiaolong and Prof. Long went to the sericulture base for the second time.

Xiaolong: Dad, can you take us to gather mulberry leaves?

Prof. Long: Okay. But it may not be what we've thought. Some methods should be learned before we begin.

David: Isn't it all about taking leaves off?

Xiaolong: It won't be that simple, I guess. Does it need any techniques?

Prof. Long: Of course. To start with, I'd like to introduce some general knowledge about silkworm rearing. Given that one kilogramme of fresh cocoons includes 500 cocoons, what do you think is the weight of the mulberry leaves consumed?

Xiaolong: It's hard to guess. Maybe 10 kilogrammes?

Prof. Long: And plus half of it.

David: 15 kilogrammes. That's a lot.

Prof. Long: Yes. Grafting is for the purpose of producing more leaves. So, some ways should be worked out to retain this during the harvest of mulberry leaves.

David: What should be done then?

Prof. Long: When picking leaves, we should consider its influence on the growth of mulberry trees as well as the needs of feeding silkworms. So, it's important to know the right methods of picking leaves.

Xiaolong: Oh, you mean we can't take them all?

Prof. Long: Partially right. The methods of picking leaves vary in different seasons.

David: I once thought leaves are only to be collected in spring.

Prof. Long: Of course not. Mulberry leaves can also be picked in summer or autumn. The harvesting methods vary based on the time when silkworms are reared and the way mulberry trees grow in each season.

David: What're the methods then?

Prof. Long: There are three basic ones, namely, leaf picking, whole shoot cutting and branch cutting.

Xiaolong: Leaf picking means selecting only the leaves; whole shoot cutting refers to collecting only the fresh shoots, right?

Prof. Long: The first method is correct but the second one is not. The shoot in the second method is an umbrella term for leaves and new buds.

Xiaolong: So this method is widely used in spring time?

Prof. Long: Yes.

David: Then, what about the branch cutting method?

Prof. Long: It means cutting branches with leaves. The method is commonly used at the turn of spring and summer.

David: It seems that gathering mulberry leaves is not an easy task.

Prof. Long: That's true. If it's done in an inappropriate way, the leaf yield will be severely reduced. The time for picking is also regulated, usually twice a day. The two suitable periods of time are from 6 a.m.

to 9 a.m. and from 4 p.m. to 6 p.m. Those who cultivate mulberry trees and raise silkworms should know what to do in different seasons with different methods. I'll find a veteran to teach you before we get started.

David: I really need to put some work into it.

Xiaolong: Me too. Let's go. It's 8:00 o'clock.

Prof. Long: OK. Let's have a try of whole shoot-cutting, the method used during spring time.

催青

采桑结束,桑蚕基地的技术员陪着大卫、小龙和龙教授来到蚕室了解养蚕过程。蚕室里面是一间间隔好的小房间。他们走进第一间,大卫和小龙看到椭圆、扁平的蚕卵密密麻麻地铺满一张张大纸。

大　卫:请问,这密密麻麻的都是蚕卵吗?

技术员:对呀,都是的。

小　龙:没想到蚕卵的颜色这么深,我还以为是白色的呢。

技术员:这些是越年卵,刚产下的蚕卵是淡黄色的,经过10天左右,就变成了这么深的颜色。

大　卫:现在就等着它们自己变成蚕宝宝吗?

技术员:蚕卵是可以自然孵化的,但那样孵化率低,而且有孵化不整齐、幼虫体质弱等问题,所以我们要帮助它们孵化,这个过程叫催青。

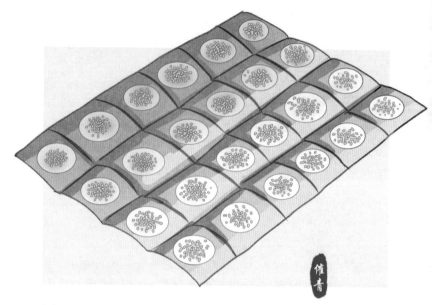

催青
Artificial Incubation of Silkworm Eggs

大　卫：催青？是人工孵化吗？

龙教授：是的，就是人工孵化，这样成活率更高。

技术员：催青是养蚕过程中很重要的一关。为了保证培育出健壮的、个头一般大的蚕宝宝，催青室里的温度、湿度、光线、空气都要控制好才行。

大　卫：看样子是标准化生产。

技术员：是的，要严格按照生产要求来。

小　　龙：难怪这个房间里放了空调和加湿器。

技术员：我们一般采用渐进催青法，随着蚕卵胚胎生长，要逐步提高室内温度，同时注意调节湿度和光线。

大　　卫：怎么调节呢？

技术员：催青的时间持续11天左右。这期间又分成三个重要阶段，也就是最长期、缩短期和点青期。最长期的胚胎体细长。你们看，这一批蚕卵正处于最长期，所以今天我们要开始给它们催青了。温度要控制在20℃左右，湿度77%，用自然光照射。

小　　龙：那缩短期和点青期呢？

技术员：接下来3天，要注意室内逐渐升温，温度控制在23℃左右。第5天，胚胎进入缩短期，需要进行高温保护和18个小时以上的连续光照，温度控制在25℃左右，湿度80%。当蚕的头部颜色变浓黑时，胚胎就进入了点青期。在点青期，胚胎即将全身变黑，这时，温度需要提升到26℃，需要昼夜遮光。1至2天后，在光照条件下，蚕卵很快就会孵化成幼蚕了。

大　　卫：从蚕卵变成蚕宝宝还真不容易呢。

技术员：是的。催青过程中的温度、湿度和光线，直接决定着蚕卵孵化率和后期成长情况，甚至会影响吐丝质量。

小　　龙：催青很有技术含量，得专门学习吧。

技术员：是啊，上岗前技术人员都需要接受专门培训。

Artificial Incubation of Silkworm Eggs

> A technician at the sericulture base brought David, Xiaolong and Prof. Long into a silkworm incubation room and told them about the process of silkworm rearing. It was a small room with many separate compartments. Walking into the room, they saw dense, ovoid and flat eggs on many sheets of paper.

David: Are they all silkworm eggs?

Technician: Right.

Xiaolong: I never thought the eggs to be incubated look so dark. I supposed they were white.

Technician: These are diapause silkworm eggs. And the newly-laid silkworm eggs are pale yellow. They will turn dark brown and become what they're now after about 10 days.

David: We'll just wait for the eggs themselves to hatch?

Technician: Eggs can hatch in the natural state, but there will be problems like low hatch rate, uneven hatch and weak larvae. Thus, human intervention is needed and it's called artificial incubation of silkworm eggs.

David: Artificial incubation of silkworm eggs? Does it mean we help silkworm eggs hatch out?

Prof. Long: Exactly. It's human-aided incubation and the survival rate is much higher.

Technician: And it's important in the process of rearing silkworms. Room temperature, humidity, light and air should be well regulated to breed healthy silkworms of similar size.

David: It sounds like standardised incubation.

Technician: Yes. It's in strict accordance with the incubation requirements.

Xiaolong: No wonder the room is equipped with an air conditioner and a humidifier.

Technician: We generally adopt gradual incubation of silkworm eggs. As the silkworm embryo develops, the room temperature is gradually increased, and humidity and light should also be under control.

David: How to regulate all these factors?

Technician: That depends. The whole process lasts about 11 days and includes three key stages, namely the longest stage, the shortening stage and the head pigmentation stage. During the first stage, the embryo looks elongated. For example, this batch of silkworm eggs is in that very stage, so today we should start artificial incubation of them. The temperature should be set around 20 ℃ with the humidity level at 77% and natural light.

Xiaolong: How about the last two stages?

Technician: In the upcoming three days, the temperature should be raised bit by bit to about 23℃. On the fifth day, the embryo enters the shortening stage, during which a higher temperature and more than 18 hours of continuous light are needed. Then, we set the temperature around 25 ℃ and the humidity level at

80%. When the head of each egg turns dark black, the embryos go into the head pigmentation stage. The black pigmentation will spread across the entire body, and at that time the temperature should increase to 26℃. Remember to block out light day and night. After one or two days, a larva will come out of each egg with the light.

David: A lot of work has been put in to help eggs become baby silkworms.

Technician: Indeed. During the artificial incubation, the three factors, namely temperature, humidity, and light, determine the hatch rate, silkworms' later development and even silk quality.

Xiaolong: Artificial incubation really involves much knowledge. That will call for special training, right?

Technician: Yes. Learning and training are required for all of us before we get started.

炕床育

> 参观完催青室,技术员带着大卫、小龙和龙教授一起进入小蚕饲育工作间去看蚕宝宝。

大　卫:哇,这些蚕宝宝好小啊。

小　龙:这是刚孵化出来的吧?

技术员:是的,这是刚催青后孵化出的幼蚕,它们需要特殊照顾。

小　龙:怎么特殊照顾呢?

技术员:幼蚕的饲养有三种方法:普通育、覆盖育和炕床育。别看这些幼蚕个头这么小,要是按单位体重算,它们的表面积是很大的,所以身体散热快、水分容易蒸发,需要在高温多湿的环境下饲养。

大　卫:具体用什么方法呢?

技术员:我们通常采用炕床育。

小　　龙：炕床？是东北人的那种大炕吗？我们南方可没有炕。

龙教授：这里有呀，不过这个炕不是给人睡觉，而是培育蚕宝宝用的。在育蚕加温过程中，我们应用了北方取暖用的炕床结构，所以这种饲育方式被称为炕床育。

技术员：对，龙教授说的没错。炕床分饲育室和地火龙两部分。

小　　龙：饲育室好懂，应该是用来养幼蚕的地方。那地火龙是什么呢？

大　　卫：要用火吗？

技术员：是的。地火龙是加热的，主要由炉灶、烟囱和多条地下烟道构成，这样饲育室的加温就会很均匀，这种间接加热的方法可以避免二氧化碳堆积。你们看，天花板和后墙上都有换气孔，用来保证室内空气流通。

小　　龙：饲育室设计上要考虑这么多因素哇。

龙教授：你们知道高温多湿、空气新鲜的环境除了有利于幼蚕生长，还有什么好处吗？

大　　卫：我猜是保持桑叶新鲜吧？

技术员：对的。高温多湿环境下，桑叶中的水分不易流失，可以长时间保鲜，这样的话，可以减少给桑次数，避免浪费桑叶。

大　卫：这样劳力和桑叶都省了。这些幼蚕什么时候能出这个屋呢？

技术员：四龄和五龄蚕就可以不用炕床育了。

加热　Heating

炕床育　*Kang*-Bed Rearing

小　　龙：怎么给蚕分年龄呢？

技术员：蚕的生长期是这么分的：催青期10—11天；幼虫期25天左右，具体是一龄期4—5天、二龄期3—4天、三龄期4天、四龄期6天、五龄期7—9天；蛹期14—18天；蛾期3—5天。

大　　卫：分得可真细啊。

技术员：是的，必须细分。一龄和二龄期，饲育室的温度得控制在27℃—28℃，湿度90%；到了三龄期，温度降1℃，湿度降到80%—85%。四龄和五龄的蚕就属于大蚕，不再需要用炕床了。

大　　卫：看来蚕宝宝的生活管理要很科学呢。

小　　龙：我有个问题，蚕龄就是按时间算吗？这中间有没有能观察到的成长现象呢？

技术员：这个问题问得好。蚕在生长过程中要经过四眠五龄。它们得不断蜕皮才能长大。

大　　卫：蜕皮？

技术员：是的，要蜕好几次皮呢。蚕每次蜕皮期间不食不动，一看就知道是休眠了。眠与眠之间就是龄。

小　　龙：那应该很容易观察到。

技术员：没错。蜕去旧皮后，蚕就进入一个新的龄期。从蚕卵孵化到第一眠是一龄期；之后醒来吃桑叶，也就是二龄期，再进入第二眠；依此类推，一直到蚕进入四龄期和五龄期，变成大蚕。

大　卫：真有意思，蚕宝宝是吃吃睡睡长大的呀。

技术员：是这样的。再过一个多月，你们可以来看蚕怎样上蔟吐丝了。

小　龙：上蔟是什么意思呢？

技术员：这个到时候我再给你们讲。

小　龙：好呀，到时候我们一定来。谢谢您。

Kang-Bed Rearing

> After they left the room of artificial incubation, the technician took David, Xiaolong and Prof. Long to the workroom for breeding baby silkworms.

David: Wow, how tiny these baby silkworms are!

Xiaolong: They're newly hatched, right?

Technician: Yes. They're just out from artificial incubation and need special care.

Xiaolong: What do you mean by special care?

Technician: Let me explain it. Normal rearing, covered rearing and *kang*-bed rearing are the three methods of baby silkworms breeding. Small as they are, these silkworms have a large body surface area per unit weight, so they can dissipate heat quite fast and dehydrate easily. Baby silkworms have to be reared in rooms with a high temperature and humidity.

David: What specific methods do you use?

Technician: We generally adopt *kang*-bed rearing.

Xiaolong: *Kang*-bed? Is it the *kang* used by people in northeastern China? There's no *kang* in the south.

Prof. Long: There actually is. The *kang*-bed in this case is not for people to sleep on, but for breeding baby silkworms. *Kang*-bed rearing owes its name to the *kang*-bed unique to the northern part of China, and the structure is readapted as a heated platform for silkworms during the breeding process.

Technician: Yes, you're right. It consists of a silkworm breeding room and *dihuolong*（地火龙）.

Xiaolong: It's easy to understand the breeding room. It's a place for raising larvae. What's *dihuolong*?

David: Is fire necessary for that?

Technician: Yes, *dihuolong*, a heated pipeline, is composed of a stove, a chimney and multiple underground flues, so that the breeding room can be heated and CO_2

buildup is avoided. Look at the ventilation holes on the ceiling and back walls. They can ensure indoor air circulation.

Xiaolong: Oh. I didn't expect so many things need to be considered when designing a breeding room.

Prof. Long: Well, it won't be easy. As we know, high temperature and humidity plus fresh air are necessary for the growth of baby silkworms. Do you know any other benefits of this condition?

David: Preserve the freshness of mulberry leaves?

Technician: Granted. In such an environment, the moisture in mulberry leaves will be lost less quickly than otherwise. The leaves can stay fresh for a longer time. In this way, the feeding times can be reduced, and there won't be any waste of mulberry leaves.

David: Both labour and mulberry leaves are saved. When can those baby silkworms leave the breeding room?

Technician: Well, not until they enter the fourth or fifth instar phase.

Xiaolong: How do we know what phases they are in?

Technician: Here're different stages of silkworms: the artificial incubation period, 10–11 days; the larva stage, around 25 days, which can be further divided into the first instar, 4–5 days, the second instar, 3–4 days, the third instar, 4 days, the fourth instar, 6 days, and the fifth instar, 7–9 days; the pupa stage, 14–18 days; and the moth stage, 3–5 days.

David: What a detailed division!

Technician: Yes. The division of different stages is really necessary. For the first to second instar silkworms, the temperature in the breeding room should be set in the range of 27 ℃ to 28 ℃ with humidity level at 90%. When they begin the third instar stage, we need to lower the temperature by 1℃ and adjust humidity to the range from 80% to 85%. The fourth to fifth instar silkworms are strong enough and no longer need the *kang*-bed.

David: It seems that the everyday life of baby silkworms is managed scientifically.

Xiaolong: But is the period of silkworms only determined by

the length of time? Are there any observable signs of growth?

Technician: Good question. Silkworms usually experience four moults and five instars during their lifetime. They have to shed their skins before entering the next developmental stage.

David: Shed their skin?

Technician: Exactly. Silkworms will moult several times. When moulting, they don't eat or move, which is called dormancy. An instar is a stage between one dormancy and the next.

Xiaolong: That would be quite noticeable.

Technician: Yes. Sloughing off old skin, silkworms begin a new instar. To be specific, silkworms enter the first instar when they sleep; after waking up to eat mulberry leaves, they fall into sleep for the second time and reach the second instar. The cycle goes on till they come into the fourth and fifth instar and become large silkworms.

David: Interesting. It turns out that to grow up into mature silkworms, all they need to do is eating and sleeping.

Technician: Yes. More than one month later, you can come and see how silkworms climb mountages and start spinning.

Xiaolong: Climb mountages? What does that mean?

Technician: I will save this question till next time.

Xiaolong: OK. We'll see you then. Thanks for your instructions today.

上蔟吐丝

> 一个月后,小龙和大卫再次来到桑蚕养殖基地观看蚕吐丝结茧。

大　卫:请问,中间这个高台子上一格一格的框架是干什么用的?里面还有蚕呢。

小　龙:应该是让蚕结茧的地方吧?

技术员:是的,这些是方格蔟,是让熟蚕结茧的工具。大蚕长成熟蚕后,会爬到蔟具上吐丝结茧。

小　龙:这是爬格子,为什么叫上蔟呢?

技术员:最原始的蔟具是用稻草、麦秆扎起来的,像山的样子,所以叫上蔟。后来为了提高蚕茧质量,增加蚕茧解舒率,人们研发了结构更好的蔟具,比如这种网格蔟具。

大　卫:解舒率是什么意思?

技术员：通俗一点讲，解舒是指抽丝时茧丝分解抽离的难易程度。解舒率高，表示茧丝容易分解抽离，出丝量就多。解舒率低，抽丝量就少，而且生丝品质也不会太好。

大　卫：明白了，这种蔟具的结构有利于蚕吐丝。

小　龙：这种方格蔟有什么优点呢？

技术员：方格蔟隔孔大小均匀，便于空气流动，能保持蔟中干

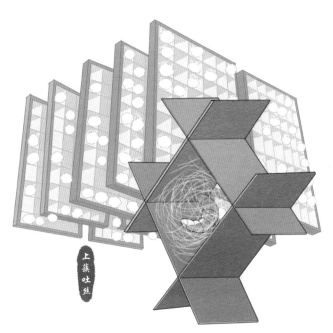

上蔟吐丝　Mounting and Spinning

燥。熟蚕可以各选其格，很好管理。

小　　龙：看，有的蚕开始吐丝了。怎么这些蚕只是抬着头晃呢？

技术员：这个动作表明马上要吐丝了。熟蚕长到一定时期就停止进食，这时它们胸部透明，抬起头部左右摆动，准备吐丝。

大　　卫：这个房间是不是也要控制温度、湿度和光线呀？

技术员：没错。从熟蚕上蔟到吐丝这一阶段，需要屋里温度适宜、干燥清洁，光线明暗均匀，这样才能保证结出高质量的蚕茧。

小　　龙：你们看，蚕用丝把自己一层一层包裹起来，这是真正的"作茧自缚"啊。

技术员：这个成语确实是这么来的。这些丝网会越来越厚，最后结成一个椭圆形的茧。

大　　卫：从吐丝到结茧需要多长时间呀？

技术员：大约3个整天，60到80小时。有的结得快些，有的慢些。

大　　卫：真有趣，蚕还像人一样，有的性子快，有的性子慢。

技术员：是的，你的比喻很恰当。

小　龙：我喜欢那句诗"春蚕到死丝方尽"。我觉得蚕很伟大，牺牲自己，造福人类。

技术员：这是文人托物言志。我倒是觉得我们中国人很聪明，不仅发现了蚕吐丝现象，而且还用蚕丝成功地织出了丝绸。

大　卫：有人说丝绸是中国人的第五大发明呢。

技术员：的确如此。蚕丝是非常优质的纺织原料，纤细轻盈，光泽度好，而且柔软顺滑，透气性好，特别适合做衣服。蚕丝还可以做蚕丝被，又轻又软，保暖又透气。

Mounting and Spinning

One month later, Xiaolong and David went to the sericulture base once again to watch silkworms spinning cocoons.

David: Excuse me. What're these grid shelves in the middle of this high table used for? There're silkworms in them.

Xiaolong: I guess it's for silkworms to spin their cocoons?

Technician: Yes. They're called grids, used for silkworms' mounting and spinning. After the large-size silkworms grow mature they will climb onto mountages to spin cocoons.

Xiaolong: It looks like they are mounting grids. Why is it called mounting?

Technician: Originally, mountages were made of straw and wheat stalks and they were quite like a mountain.

Mature silkworms were observed climbing up them, so it was called mounting. Later, better structured devices like grids were developed to improve cocoon quality and increase reelability.

David: What's reelability?

Technician: It refers to the level of difficulty for reeling silk from cocoons. A high reelability indicates much more silk can be easily drawn from cocoons while a low one shows undesirable reeling efficiency. If the reelability is low, the raw silk will be of bad quality.

David: I see. The grid shelves are structurally adjusted to help silkworms extract silk.

Xiaolong: What are their advantages?

Technician: The grids are uniform in size and well-ventilated, so they can stay dry. Mature silkworms can choose their respective grids and they're easy for management.

Xiaolong: Look! Some silkworms start spinning. Why are others just shaking their heads?

Technician: This is a sign that they're going to start spinning. Mature silkworms stop eating as their chests grow transparent. Afterwards, they will lift their heads to shake from side to side, ready for spinning.

David: The temperature, humidity and light of the room should also be regulated, right?

Technician: That's for certain. From mounting to spinning, moderate temperature, dry and clean environment as well as medium light are a must to ensure high quality cocoons.

Xiaolong: Look, the silkworms are wrapping themselves with their own silk layer by layer, which literally illustrates the Chinese idiom "to spin a cocoon to confine oneself".

Technician: This is exactly where the idiom comes from. And the web of silk will become thicker until it becomes an oval cocoon.

David: How long will it take to spin a cocoon?

Technician: Around 3 days, or 60 to 80 hours. The speed they weave also differs individually.

David: Interesting. Silkworms are like people. Some enjoy quick pace while others don't.

Technician: Exactly. You sum it up aptly.

Xiaolong: I like the line: "Spring silkworm till its death spins silk." Silkworms are great creatures, sacrificing themselves for the benefits of humans.

Technician: That's right. It's the literati projecting their aspirations onto silkworms. Personally, I think the ancient Chinese were smart, because they not only discovered that silkworms could spin cocoons, but also found ways to weave silk fabrics with them.

David: Some people even hail silk fabrics as the fifth great invention of China.

Technician: It's true. Silk is a unique type of quality textile material. Light yet strong, it's glossy, smooth and breathable, perfect for making clothes and silk quilts.

煮茧缫丝

> 参观完簇室,大卫和小龙来到缫丝坊,这是一个仿照传统缫丝工艺设立的工作坊。技术员给他们讲解了缫丝的知识和整个技术流程。

技术员:我们刚才看了桑蚕吐丝结茧,现在再来了解一下蚕茧是怎么变成丝线的,这个行话叫缫丝。我们就从剥茧、选茧开始吧。

大　卫:剥茧?就像剥洋葱那样把蚕茧剥开?

技术员:有点儿类似,就是将蚕茧表层的一些杂乱、松散的丝缕剥去。剥茧后要对蚕茧进行分类,这叫选茧。最好的茧叫上茧,缫出的生丝可用来制作高级丝织品;最差的茧叫下茧,不能用来缫丝。

大　卫:怎样分辨茧的好坏呢?

技术员:好茧的茧层厚实,茧形整齐,颜色洁白均匀,表面没有疵点。差茧的问题各不相同,比如有的表面发霉、有的有大面积油污，或者茧形畸形、个头太小。

小　龙：那选茧之后就该煮茧了吧？

技术员：是的。干茧中含有大量的丝胶，因为过于紧实，无法用来缫丝。煮茧的目的就是利用水和热的作用，使蚕茧中的丝胶膨润，降低黏着力，这样才能抽出丝来。

大　卫：怎么煮呢？

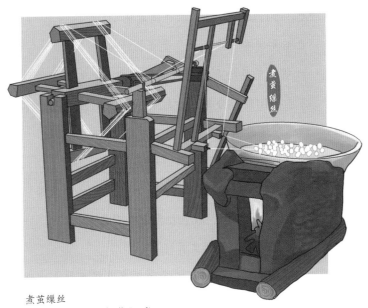

煮茧缫丝
Cocoon Boiling and Silk Reeling

技术员：煮茧不是把茧一股脑儿倒在一起煮，要分好几步。首先，将干茧放入容器里浸泡2到3个小时，使表面形成一层水膜，水温控制在50℃到70℃之间，防止高温把蚕茧煮烂。

大　卫：看来还得好好掌控温度呢。

技术员：是的。泡茧的同时需要调配化学药剂，增强液体的渗透力。将浸泡后的蚕茧、化学药剂和60℃左右的水一同放入真空容器里，蚕茧充分渗透后，解舒率会提升。这样缫丝效率会更高，生丝的产出量也更大。渗透之后，就可以开始煮茧了。

小　龙：煮茧要用多少度的水？煮多长时间呢？

技术员：煮茧最重要的是控制时间和温度，45℃左右最佳，通常要10到15分钟。煮过的茧放入温度更高的水中调整一段时间，才能进行下一步工作——索绪。

小　龙：索绪就是找到蚕丝的头，对吗？

技术员：是的。你们看，索绪需要这个工具——索绪帚。用它摩擦蚕茧表面，就能把蚕丝头引出来。

大　卫：哦，这还有专门的工具呢。

小　龙：古时候人们能够把蚕茧和织布穿衣联系起来，也真是太聪明了。

煮茧缫丝　Cocoon Boiling and Silk Reeling　095

索绪
Finding the Ends of Silk

技术员：没错，古人很有智慧。关于缫丝还有很多传说呢。

大　卫：有什么有趣的故事？快给我们讲讲吧。

技术员：有一个嫘祖怎么找到茧的丝头的传说。嫘祖是黄帝①的妻子。相传有一次黄帝打了胜仗，为了庆祝胜利，蚕神化作美女，献上了洁白的丝。黄帝特别喜欢这个礼物，就让嫘祖教人栽桑养蚕，但很长时间没有人能弄清楚怎样才能把蚕茧制成丝。

大　卫：后来呢？

技术员：嫘祖得到蚕茧后，也想不出来怎样抽丝。

小　龙：嫘祖不是蚕神吗？

技术员：别着急，听我讲她是怎么成为蚕神的。一天，她坐在桑园里喝水，手里拿着蚕茧仔细琢磨。不小心手一滑，蚕茧落入了热水中。嫘祖着急把蚕茧从热水中取出来，就拿手边的筷子去捞。你们猜发生了什么？

大　卫：用筷子找到了丝头？

技术员：是的。嫘祖惊讶地发现，泡了热水的蚕茧出了丝。她用筷子轻轻一挑蚕茧，蚕丝居然连续不断地从茧中拉了出来。最初的索绪工具就是仿照筷子设计的两个木签。

小　龙：哦，嫘祖成为蚕神应该是筷子立的大功。

技术员：是的。为了提高效率，就有了索绪帚这个工具，这么多竹签聚在一起，一次能引出好多根丝。

小　龙：那缫丝后是什么工序呢？

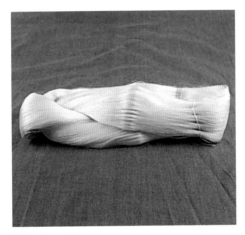

生丝　Raw Silk

技术员：是复摇和整理，最后绞丝变成成品生丝。

大　卫：复摇是什么？

技术员：就是用复摇机除去糙片、杂质，整理好接头，提高丝的洁净度。然后把丝线从复摇机上取下，逐片绞好打包，就是成品的生丝。

> 看到工作间的展台上摆放的亮灿灿的成品生丝，大卫和小龙忍不住伸手去摸了摸。

小　　龙：光泽度真好，摸着好顺滑呀。那下一步就应该是染色了吧？

技术员：是的。天然蚕丝是白色的。练染就是给蚕丝染色。

大　　卫：给蚕丝染色有什么讲究呢？

技术员：中国的练染技术起源很早。早在2000多年前的周朝，政府就设"染人"之职，专门负责染色的各个环节。春天煮练曝晒丝帛，夏天染黄、红和浅黑色，秋天染其他颜色，冬天进献染好的丝帛成品。

小　　龙：2000多年前人们就已经能染丝了，真是厉害。

技术员：是的。在我们湖州也能找到古代织染业活动的依据。你们知道这附近有条小河叫墨浪河吗？

小　　龙：知道，好像是染坊很多，把水染成了黑色。

技术员：正是如此。可见古代这里丝织业是非常发达的。

大　　卫：那现在河水还很黑吗？

技术员：哈哈，现在当然不黑了，但名字保留了下来。

小　　龙：我觉得这个名字挺有趣的，还很有诗意呢。

技术员：现在缫丝厂都是自动化机械操作了，也都非常注意环保，不会再出现污染河流的情况了。我们建了这个简易手工缫丝坊，为的是让大家了解传统的缫丝染色技艺。

小　龙：谢谢您带我们参观，还耐心地给我们这两个外行讲解。

大　卫：谢谢您。

注释：
① 黄帝：中国古代部落联盟首领，五帝之首，史称"人文初祖"。

Cocoon Boiling and Silk Reeling

Leaving the cocoonery, David and Xiaolong went to the workshop on traditional silk reeling. The technician explained to them the concept of silk reeling and the whole process involved.

Technician: We have just learnt about how silkworms produce silk and spin cocoons. And now we are going to watch how to turn cocoons into silk threads, or the silk reeling. Let's start with cocoon stripping and sorting.

David: Cocoon stripping? Is it like peeling the garlic?

Technician: Kind of. It involves removing the rough and loose silk fibres from the outer layers of cocoons. Then it's time for cocoon sorting, which is to classify cocoons into high-grade and low-grade ones. The raw silk reeled from the former can be used to make premium silk fabrics, while the latter are unqualified for silk reeling.

David: But how to tell good cocoons from defective ones?

Technician: For superior cocoons, the shape is neat; the layer is thick and of uniform size; the colour is evenly and purely white with no flaws on the surface. For inferior cocoons, they have defects such as mouldy surface, oil stains at the outer layer, irregular shape or small size.

Xiaolong: So, the next step is cocoon boiling?

Technician: Yes. Dried cocoons contain a large amount of sericin, which makes them too firm to be reeled. That's why we boil cocoons. After the swelling and softening of the sericin in cocoons, the stickiness can be reduced. Then silk thread can be drawn out easily.

David: How do people boil them?

Technician: We are not supposed to boil them altogether. The boiling actually takes several steps. First, we soak dried cocoons in a container for 2 to 3 hours, so that a film of water is formed on the surface. Then we keep the temperature at about 50°C to 70°C to prevent cocoons from being over-boiled during the steaming process.

David: It seems that we have to hold a desired temperature.

Technician: Exactly. Then, we blend a chemical to enhance the permeability of water. Next, we put soaked cocoons, the chemical and water at about 60°C into a vacuum container until cocoons are fully immersed. This can help increase the reelability and promote the efficiency of boiling, and hence greater output. After that, we can start boiling cocoons.

Xiaolong: What temperature of the water should be? And how long is the boiling process?

Technician: The priority is to have a good command of the time and temperature. 45°C and 10–15 minutes would be perfect. Then we put boiled cocoons into hotter water before the next step: to grope them.

Xiaolong: I guess it means finding out the ends of silk threads, right?

Technician: Sure. Look, this is a groping brush, which is used to rub cocoon surface to ravel out the ends of silk.

David: Oh, a specialised tool is needed here.

Xiaolong: It's such a clever feat that the ancient Chinese used silkworm cocoons for weaving clothing.

Technician: Exactly. They were very wise. A lot of legends are related to silk reeling.

David: Is there any interesting story? I can't wait to hear more about it.

Technician: There's a story about how Lady Leizu, the wife of Yellow Emperor[1], found the ends of silk. Legend has it that after Yellow Emperor won a battle and celebrated the victory, the Silkworm Goddess, disguised as a beauty, offered pure white silk for the occasion. Yellow Emperor was very happy about it and asked Leizu to teach people how to plant mulberry trees and raise silkworms so that they could have silk. But for a long time, nobody could figure out how to turn cocoons into silk threads.

David: What happened then?

Technician: Lady Leizu couldn't figure out how to extract silk for the moment, either.

Xiaolong: Isn't Lady Leizu the Silkworm Goddess?

Technician: Sure… But let me tell you how she became the Goddess. One day, she was drinking a cup of hot tea at the mulberry garden while observing a cocoon in her hand. Suddenly, the cocoon slipped through her fingers and fell into her cup. Hurrying to fish it out, Leizu grabbed the chopsticks at hand. Guess what happened?

David: She found the ends of silk with the chopsticks?

Technician: That's right. Leizu accidentally discovered that the silk threads were continuously pulled out from the soaked cocoon after picking an end with her chopsticks. So those early groping brushes were made of two simple wooden sticks, which was exactly inspired by a pair of chopsticks.

Xiaolong: Wow, much credit should be given to the chopsticks for helping Leizu become the Silkworm Goddess.

Technician: Definitely. Later, the groping brush was invented to improve the efficiency, which could pull out many strands of silk at one time.

Xiaolong: What're the following steps after reeling?

Technician: To get the ready-to-use silk, the remaining steps are re-reeling, lacing and skeining.

David: What's re-reeling?

Technician: It refers to the step of removing rough pieces and impurities and sorting out the ends of skeins on re-reeling machines so as to make the raw silk cleaner. Then we take re-reeled silk off the machine to wind it into skeins and pack them. That's how we get the processed silk.

> Seeing the shiny raw silk on the booth in the workshop, David and Xiaolong couldn't help but touch it.

Xiaolong: It's glossy and smooth. What's the next? Dyeing?

Technician: Right. Natural silk is white. We need to colour it.

David: What's special about dyeing?

Technician: China's dyeing techniques date far back. As early as more than 2,000 years ago, during the Zhou Dynasty, a special position for dyeing was set up, and it was named the Dyer. He was responsible for the whole dyeing process. In spring, silk was boiled and exposed to the sun; in summer, it was dyed in yellow, red or light black; in autumn, other colours were dyed; and in winter, the dyed silk was offered to the ruler.

Xiaolong: Awesome! Ancient Chinese were already able to dye silk over 2,000 years ago.

Technician: Indeed. Weaving and dyeing in ancient China can also be traced back to Huzhou. Did you know that there was a river nearby, called the Ink River?

Xiaolong: Yes. Dyeing houses used to be found all over there, so the water was dyed inkish.

Technician: Exactly. The textile industry in Huzhou was well developed in ancient times.

David: Is the river still blackish?

Technician: No more. But the name has been preserved.

Xiaolong: Aha, quite an interesting and poetic name.

Technician: Well, that's history. Today, reeling mills are operated by machines and people are quite environmentally conscious, so the pollution of rivers is a thing of the past. And this simple and hand reeling workshop we've built is only for the purpose of introducing the traditional silk reeling and dyeing techniques to visitors.

Xiaolong: Thank you for showing us around and explaining everything.

David: Thanks a lot.

Note:

1. Yellow Emperor: A leader of tribe alliances and the first of China's Five Emperors in legend. He is honoured as the primogenitor of the Chinese nation.

丝绸

> 参观完桑蚕养殖基地，小龙和大卫回到小龙家。吃完晚饭两人又跟龙教授聊起了丝绸的话题。

龙教授：中国的丝绸举世闻名。古时西方人对中国的称呼还和丝绸有关呢。他们称中国Seres，意思是"丝的国度"。

大　卫：这个我听说过的。

小　龙：还挺有道理的。丝绸就是中国人发明的嘛。

龙教授：丝绸和别的东西不一样，古代的实物很难保存下来，所以丝织文物都特别珍贵。

小　龙：我知道世界上现存年代最久、最轻最薄的衣服是丝织衣服，是在长沙发现的。

龙教授：是的，是1972年长沙出土的素纱襌（dān）衣，是2200年前汉代的丝织衣服。

小　　龙：我去长沙参观博物馆的时候看到过素纱襌衣，还以为那个字念禅（chán）呢。

龙教授：纱是古代丝绸中出现最早的品种之一，是用单经单纬的丝织成的。

小　　龙："素"是冷色的意思吗？

龙教授：不是的。丝绸从染色和花纹上可以分为素和花两种。"素"是说丝只经过练、漂、染后直接编织，上面没有任何图案。"花"是指织上或者印上花纹图案。

大　　卫：素纱襌衣是什么样的衣服呢？

龙教授：襌衣指的是没有里子的衣服，实际上就是我们平常说的单衣。长沙出土了两件素纱襌衣，都非常轻薄，其中一件长1.6米，重量只有48克。

大　　卫：这么轻呀，可真神奇。

龙教授：你们看，汉代缫丝纺织的技术是不是很高超呀？

小　　龙：是呀。现在年代最久的丝绸在南方发现，是不是说明最早产自南方？

龙教授：还真不是这样。应该说，1000年前，北方生产丝绸更多，丝绸作坊遍布整个黄河流域。南宋以后南方的丝织业才逐渐发展起来。

大　卫：为什么后来丝织业会转移到南方呢？

龙教授：主要原因是政治经济中心南移，南宋管辖的地区只有长江、淮河以南的地区。还有一个原因是气候，在这之前很长一段时间，北方气候比较温和湿润，很适合种桑养蚕。后来气候发生了变化，北方变得寒冷干燥，种桑养蚕也就自然转移到南方地区了。

大　卫：原来历史上气候也有变化呀？

龙教授：是的，大起大落的变化有很多次，而且对中国南北方农业的改变有很大的影响。丝绸自古就是中国特有的纺织品，不仅我们中国人自己喜欢，在其他国家也很受欢迎。

大　卫：历史上，中国的丝绸、茶叶和瓷器在欧洲都很受欢迎。不是有个"丝绸之路"吗？

龙教授：是的。古代的"丝绸之路"是从长安经甘肃、新疆至中亚、西亚、地中海各国的一条陆上贸易通道，其中丝绸是主要贸易物品之一。

小　龙：好像丝绸和茶叶一直都是我国古代出口的主要产品。听说因为中国丝绸的大量入境，还造成罗马帝国的黄金大量外流，是吗？

龙教授：是的，真是这样。在古罗马，不仅皇帝和贵族喜

　　　　　欢穿中国丝绸，连老百姓都跟着买，造成大量黄金外流，因此罗马帝国制定了相关法令，禁止人们穿丝绸。

大　　卫：丝绸这么受欢迎是因为花色漂亮吧？

龙教授：不完全是，主要和丝绸面料本身有关。桑蚕丝是人类最早利用的动物纤维之一，光滑轻盈、透气性好，被称为"纤维皇后"。穿丝绸衣服既透气又光滑，对皮肤还有保护作用。

大　　卫：丝绸衣服穿起来确实很舒服。

小　　龙：我们生产丝绸这么多年，品种一定很多吧？过去一说谁穿得好，就会用"绫罗绸缎"这个词。

龙教授：是的，绫罗绸缎其实是四种不同的丝织品。

小　　龙：那不同的丝绸有什么不同之处呢？我只知道都是丝绸。

龙教授：丝绸是个统称，根据织物的组织结构、加工工艺和绸面花纹，可以分成14大类、35小类。

大　　卫：这么多品种呀。

龙教授：是的。素纱襌衣用的纱就是14大类中的一种。我们再说说"绫"。最早的绫表面呈现叠山形斜路，看上去就像是冰凌的纹理，所以叫绫。传统绫是一种

　　　　　　暗光织物，质地轻薄、柔软。绫最早出现在魏晋时期，在唐朝发展到巅峰。唐代官员的官服布料用的就是绫，不同级别用的颜色不同。

大　卫：唐朝的官服在颜色上怎么区分呢？

龙教授：三品以上官员穿紫色，五品以上朱色，六品黄色，七品绿色。

小　龙：绫罗绸缎这四种丝绸中，哪一种出现得最早？

龙教授：罗出现得最早。据记载，2500年前就有罗这种织物。它发展的时间很长，到1000多年前的宋代达到顶峰，有很多品种，其中最有名的是杭罗。

大　卫：杭罗的意思就是杭州出产的罗吗？

龙教授：是的。罗这种织物轻薄纤细，非常透气，适合做夏天的衣服。

大　卫：绸和缎是什么时候出现的？

龙教授：绸最早出现在2000多年前的西汉时期，它质地紧密，是丝织品中最重要的一类。四种丝织物里最年轻的是缎，目前还没发现早于宋代的缎类织物。缎的特点是平滑光亮、质地柔软、色彩丰富、纹路精细。以前人们习惯把绸缎作为丝织物的代称，现在是用丝绸代称。

小　　龙：所以过去出售丝绸的商店叫绸缎庄，现在的叫丝绸商店。

龙教授：是的。

大　　卫：那现在中国人还穿绫罗绸缎吗？

龙教授：绸和缎还是很常见的，一般用于制衣。绫虽然软滑，但质地不牢，现在多用来装裱书画。杭罗现在只有一家工厂生产，产量很少，已经被列入《非物质文化遗产名录》，其他品种的罗已不多见。

大　　卫：多可惜呀，不能恢复了吗？

龙教授：到现在为止还不能，罗的织造工艺非常复杂。

小　　龙：爸爸，我看中国丝绸博物馆发布公告，说他们有个缂丝展。缂丝是什么呀？

龙教授：缂丝属于丝绸35小类中的一类，是很特别的一种。

小　　龙：那我们预约去参观一下，爸爸，您和我俩一起去，给我们讲讲好不好？

龙教授：行，没问题。

大　　卫：那太好了，我来预约。

Silk Fabrics

> After a rewarding visit to the sericulture base, Xiaolong and David went back to Xiaolong's home. They brought up the topic of silk fabrics and talked about it with Prof. Long after dinner.

Prof. Long: China boasts world-famous silk fabrics. In ancient times, China was known to westerners as "Seres", which means "the Land of Silk".

David: I heard of it.

Xiaolong: It makes sense. After all, the silk fabric is an invention of the Chinese people.

Prof. Long: Different from other materials, silk is hard to preserve. That's why silk relics are of great value.

Xiaolong: I knew that the world's earliest, lightest and thinnest surviving garment was made of silk and found in Changsha, in the south of China.

Prof. Long: Yes. It's the *susha danyi* (素纱襌衣), or plain-yarned unlined gauze garment unearthed in 1972, a textile product back in the Han Dynasty, which dates back to some 2,200 years ago.

Xiaolong: I've seen the name when I visited Hunan Museum in Changsha. The third character *dan* (襌) looks strange. I thought it was *chan* (禅).

Prof. Long: Plain gauze, among the earliest types of silk fabrics in ancient times, was made by interwoven single warp and weft threads.

Xiaolong: Does the word "plain" mean the fabric in cool colours?

Prof. Long: No. Silk fabrics can be divided into plain and flowery ones by colour and pattern. For the former, after being boiled, bleached and dyed, it is woven without any patterns. For the latter, it is woven or printed with various patterns.

David: What kind of clothes is the plain-yarned unlined gauze garment? How was it woven?

Prof. Long: *Danyi* (襌衣) refers to a garment without a lining.

There are two unlined gauze garments excavated in Changsha. Both of them are thin and light, and one of them is 1.6 metres long and weighs only 48 grammes.

David: Amazing! It's so light.

Prof. Long: It proves that the silk reeling and weaving techniques were already quite advanced during the Han Dynasty.

Xiaolong: Yes. It's no surprise that the garment was found in southern China. Does it indicate that silk was originally produced in southern China, right?

Prof. Long: Not really. 1,000 years ago, more silk fabrics were produced in northern China, with workshops all over the Yellow River Basin. But after the Southern Song Dynasty, the weaving industry in southern China gradually developed.

David: Why did the centre of textile industry move southward?

Prof. Long: Mainly because of the southward relocation of the political and economic centre. The Southern Song Dynasty governed the areas of the south of the

Yangtze River and the Huaihe River. Climate is another factor. For a long time, the climate in the north was mild and moist, which was more suitable for mulberry cultivation and silkworm rearing. Later, the climate became cold and dry, and the southward relocation began.

David: There was a climate change in history?

Prof. Long: Right. In fact, there were quite some in history. Each had a great impact on the agricultural modes in northern and southern China. Silk fabrics have been unique to China since long ago and they enjoyed popularity both domestically and internationally.

David: Historically, Chinese silk fabrics, tea and porcelain were desired by many Europeans. And there was a well-known Silk Road, right?

Prof. Long: Yes. The Silk Road is an overland trade route connecting Chang'an (today's Xi'an), Gansu and Xinjiang with Central Asia, West Asia and the Mediterranean countries. And silk fabrics were one of the major trade items back then.

Xiaolong: It seems that silk fabrics and tea had always been the major exports. I've heard that the Roman Empire had imported such large quantities of Chinese silk fabrics that the trade triggered an outflow of gold. Is it true?

Prof. Long: That's true. In ancient Rome, not only emperors and nobles, but also the general public followed the trend to buy and wear silk clothes. Due to the big outflow of gold, the Roman Empire issued a decree that forbade people from wearing silk garments.

David: Is silk so popular because of its beautiful colours and patterns?

Prof. Long: Not exactly. Silk fabric as a unique material has a say. The silk is one of the earliest animal fibres produced by humans. Smooth, lightweight and breathable, it's crowned as the "Fibre Queen" and does good to our skin.

David: They are really comfortable to wear.

Xiaolong: As a long-time producer of silk fabrics, China must have various types of them, right? The four characters in Chinese *ling*（绫）, *luo*（罗）, *chou*（绸）,

duan（缎）were often used collectively to describe luxury clothes.

Prof. Long: You're right. Each refers to a kind of textile, namely twill damask, gauze, tabby silk fabric and satin.

Xiaolong: But what's the difference among these types? I just know they are all silk fabrics.

Prof. Long: Silk fabric is an umbrella term. Based on the weaving structure, processing technology and pattern, it falls into 14 categories and 35 subcategories.

David: Wow. That's a lot.

Prof. Long: Indeed. Let's talk about *ling*, or twill damask. The earliest *ling* had mountain-shaped diagonal stripes on its surface, which looked like the crystalline structure of ice (凌, or *ling*). This is how *ling* got its name. Traditionally, *ling* is dim-coloured, thin and soft. It first appeared in the Wei and Jin dynasties some 1,400 years ago, and developed to its peak during the Tang Dynasty 300 years later. The robes of Tang officials were made of *ling*, but their colours differed by ranks of course.

David: How to differentiate the rank by the colour of robes?

Prof. Long: Government officials from and above Rank 3 were allowed to wear purple, and those from and above Rank 5 wear vermilion, Rank 6, yellow, and Rank 7, green.

Xiaolong: Among the four kinds of silk fabrics, namely *ling*, *luo*, *chou* and *duan*, which one is the oldest?

Prof. Long: That would be *luo*, or gauze, which could be traced back to some 2,500 years ago. Over time, it reached its peak during the Song Dynasty around 1,000 years ago. There're many varieties with *Hangluo* gauze as the most famous.

David: So it's produced in Hangzhou?

Prof. Long: Yes. Woven by thin filaments, *luo* is light and breathable. It's perfect for making summer clothes.

David: What about tabby silk fabric and satin?

Prof. Long: Tabby silk fabric, or *chou* in Chinese, first appeared during the Western Han Dynasty some 2,000 years ago, and was the most important type of textile with a close texture. For satin, or *duan* in Chinese, the youngest among the four, no trace has been found prior to the Song Dynasty. It's smooth, shiny and soft with coloured and fine patterns. People used *chou* and *duan* to refer to silk fabrics in Chinese, but today we use *si* (silk) and *chou* to describe them.

Xiaolong: No wonder shops dealing in silk fabrics were called *chouduan* shops but nowadays they are *sichou* shops.

Prof. Long: You're right. This is a good explanation of the different names.

David: Do people today wear clothes made by any of the four textiles?

Prof. Long: Tabby silk fabric and satin are still common, which are usually used for making clothes. Smooth and soft as twill damask is, it doesn't have a firm texture, so it's mostly used for framing paintings and calligraphy works. As for gauze, only *Hangluo* gauze is produced on a small scale because there's only one factory left. And it has been inscribed on the List of the Intangible Cultural Heritage of Humanity in China. Other kinds of gauze have gradually faded from public view.

David: What a pity! Could other varieties of gauze be recovered?

Prof. Long: No, at least not for now. The weaving techniques are said to be quite sophisticated.

Xiaolong: Dad, China National Silk Museum posted an announcement that a Kesi exhibition is around the corner. What's Kesi?

Prof. Long: Kesi is one of the 35 subcategories of silk fabrics I mentioned just now. It's a special textile.

Xiaolong: Why not book an appointment to visit it? Dad, could you come with us and tell us more about it?

Prof. Long: No problem.

David: Hooray. I'll make an appointment.

缂丝

> 龙教授带着小龙和大卫一起去中国丝绸博物馆看缂丝展，感受丝织品的艺术魅力。

小　龙：爸爸，这就是缂丝吗？看这个，正反两面的图案一模一样，很像苏绣的双面绣呀。

大　卫：感觉缂丝和苏绣不太一样，可是我说不出来什么地方不一样。缂丝是不是比苏绣更厚些？

龙教授：是厚些。这是两种完全不一样的丝织品。苏绣是绣出来的，缂丝是织出来的。不过它们的工艺流程中有一步是相同的，就是都要做劈丝。

小　龙：劈丝，就是把丝线再次细分吗？

龙教授：是的。如果需要的话，劈丝后还要把几种颜色的线再拼在一起，这叫"和花线"。

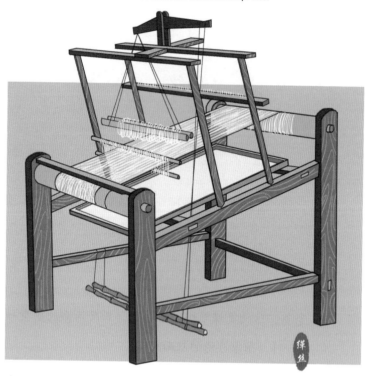

缂丝　Kesi

大　卫：缂丝是不是比其他丝织品更难织呀？

龙教授：缂丝的织法比较复杂。但无论做什么缂丝品，都少不了结、掼、勾、戗①四个基本技法。缂丝在宋代发展到鼎盛。缂丝织造时以小梭织纬，根据纹样多次变换色丝。成品只露纬丝不露经丝。

小　龙："缂丝"是什么意思？

龙教授：缂丝也叫刻丝，是一种有双面立体感的丝织工艺品。它挑经显纬，形成花纹边界，装饰性很强，像雕刻一样有立体感，有人称它是"雕刻的丝绸"。因为用工费时，所以有"一寸缂丝一寸金"的说法。

大　卫：这么珍贵呀？

龙教授：是啊。你们来看看这幅缂丝画，注意它线条的勾勒和色彩的搭配。

小　龙：纹路很细密。我也喜欢它的色彩，真是雅致漂亮。

大　卫：我很好奇，这种立体感是怎么织出来的？

龙教授：主要是用了缂丝工艺中最特别的通经断纬法。这种技艺不仅可以织出各种花纹，还可以呈现出极强的立体感。

大　卫：那缂丝只是用来制作艺术品吗？

龙教授：宋朝以前主要是制作生活用品。宋朝有一个很特别的现象，就是雅文化非常发达。缂丝本身有一定的艺术性，所以很自然就转向织造艺术品了。你们看对面展柜里就是生活用品。

小　龙：哇，用缂丝作靴子，太奢侈了吧。

龙教授：这些是700年前元代的东西，当然是贵族和有钱人才穿得起，不过的确华贵。用缂丝做服装和饰品主要是在元代。

大　卫：缂丝最早出现在什么时候？

龙教授：目前，中国发现最早的缂丝实物，是1973年在新疆出土的一件缂丝腰带，据考证是公元7世纪隋唐时期的物品。

小　龙：那也有1300多年了。爸爸，这块长方形的缂丝是做什么用的？

龙教授：这是明代的缂丝佛经盖幅，你们看，中心这块织的是八宝图案②。缂丝也常用于制作宗教物品。

大　卫：缂丝制品的用途很广泛。

龙教授：是的。缂丝有一个优点，就是强度远远高于其他丝绸类，所以在历代存留至今的丝绸品中，缂丝保存得最完好。缂丝被誉为"丝中之圣"，代表了中国高超的传统织造工艺，蕴含了独特的中国文化内涵。

注释：

① 结、掼、勾、戗：缂丝的四种基本技法。

② 八宝图案：佛教图案，指法轮、法螺、宝伞、白盖、莲花、宝瓶、金鱼、盘长八种图案。

Kesi

> Prof. Long took Xiaolong and David to visit the Kesi exhibition and appreciate the art of silk weaving in China National Silk Museum.

Xiaolong: Dad, is this a work of Kesi? The patterns on both sides are identical, just like the double-sided Suzhou embroidery.

David: I think they differ a lot, but I can't tell the difference. Is Kesi thicker than Suzhou embroidery?

Prof. Long: A little bit. But they are totally different. Unlike Suzhou embroidery, Kesi is something woven. However, the two indeed have one common step in their production: splitting silk threads.

Xiaolong: Does it mean splitting a single silk strand into finer threads?

Prof. Long: Exactly. After the splitting, the thread of various colours may be twisted together, which is called mixing colours.

David: Compared with Suzhou embroidery, Kesi is more sophisticated, right?

Prof. Long: Yes. There're four basic techniques of every Kesi work, namely *jie*(结), *guan*(掼), *gou*(勾), *qiang*(戗)[1] in Chinese. Kesi reached its peak during the Song Dynasty. Small shuttles are used to weave the weft, which are changed with the colour of the patterns. Only the weft threads are visible in the finished piece.

Xiaolong: What does Kesi mean exactly?

Prof. Long: Kesi is a double-sided weaving craft with a three-dimensional effect. With a weft-faced structure, the pattern is woven like something engraved, so this highly decorative art is also called engraved silk. It's time-consuming and labour-intensive, thus there is a saying: "An inch of Kesi is worth an ounce of gold."

David: Wow. It's so precious.

Prof. Long: Yes. Look at this Kesi painting. Pay attention to its lines and combination of colours.

Xiaolong: It has a fine texture. I like the colour and it looks elegant and beautiful.

David: I am curious about how to produce such three-dimensional effect.

Prof. Long: Then we have to speak of one of the most important techniques of Kesi weaving, the technique of continuous warp and discontinuous weft. It enables the presentation of various patterns in a three-dimensional way.

David: Is Kesi only used for making works of art?

Prof. Long: Actually, prior to the Song Dynasty, it mainly served as the material for household items. However, Kesi presents its artistic quality cherished in the culture of the Song Dynasty, hence the emergence of Kesi artworks. Look, the showcase opposite displays household Kesi goods.

Xiaolong: Wow, it's so wasteful to wear Kesi boots.

Prof. Long: Well, that's true. Back to the Yuan Dynasty, some 700 years ago, only the noblemen or other rich people could afford them and they indeed looked good and dignified. Clothes and accessories woven by Kesi mainly came from this period.

David: When did Kesi first show up?

Prof. Long: The earliest piece we know is a Kesi belt discovered in Xinjiang back in 1973. It is believed to be from the Sui and Tang dynasties, around the 7th century.

Xiaolong: That's more than 1,300 years old. Dad, what's this rectangular piece of Kesi used for?

Prof. Long: It's a cover for Buddhist scriptures from the Ming Dynasty. The centre features the eight Buddhist treasures[2]. A lot of other religious items are also found to be made of this textile.

David: It turns out that Kesi finds its application in many areas.

Prof. Long: Right. Kesi is much stronger than any other silk handicrafts. Therefore, Kesi is the best-preserved textile among all the silk pieces that survived to this day. Kesi is crowned as "the King of Silk Fabrics". It demonstrates the superb craftsmanship of the traditional Chinese silk weaving and the uniqueness of Chinese culture.

Notes:

1. *Jie, guan, gou, qiang*: They're the four basic techniques used in weaving Kesi.

2. Eight Buddhist treasures: They refer to eight Buddhist patterns, including a wheel (the symbol of Buddhist doctrines, which leads to perfection), a conch shell (the symbol of victory), a parasol (the symbol of the Buddha as the universal spiritual monarch), a victory banner (the symbol of attainment of enlightenment), a lotus flower (the symbol of purity), a treasure vase (the symbol of the elixir of life or fulfillment of all wishes), a pair of fish (the symbol of freedom from restraint) and a knot of eternity (the symbol of longevity).

缂丝画和缂丝扇面

小　龙：大卫，快来看这幅画，画面色彩多柔和呀。还有那两只小鸭子，活灵活现的，真可爱。

龙教授：这是典型的宋朝风格，典雅含蓄，是宋朝女缂丝工艺家朱克柔的名作《莲塘乳鸭图》。

小　龙：看看，这蜻蜓、小鸟、一对白鹭，真的很生动。还有荷花和荷叶，多精细呀，荷叶的纹理都清清楚楚。

大　卫：这么精致的画面，织起来肯定很费功夫。

龙教授：是的。据说用了200多种丝线，光是太湖石的蓝色就有25种。据说缂织这幅画用了8年的时间。

小　龙：我想到一个问题。织缂丝画的人还需要有很好的绘画基础吧？

大　卫：或者是不是可以请别人先画好底稿，然后再进行缂丝创作？

缂丝画和缂丝扇面　Kesi Paintings and Fans　133

缂丝画　Kesi Painting

缂丝扇面　Kesi Fan

龙教授：通常是大卫说的这样，这种做法也叫摹刻。但朱克柔是自己画，据说她很擅长绘画。她的缂丝作品题材广泛，人物、花鸟都是她创作的对象。她的作品现存7件，其中这件《莲塘乳鸭图》收藏于上海博物馆，是国家一级文物，难得能看到它在此展出。

小　龙：爸爸，这幅缂丝画上有好多人物，织起来难度很大吧。

龙教授：应该是特别难。这是明朝的作品《瑶池献寿图》，描绘的是神话传说中西王母瑶池庆寿的故事，画上人物都是神仙。

大　卫：这些神仙神态都不一样，真有意思。

龙教授：你们再看这件作品，里面用了一些特别的缂丝技术。比如在边缘用到了勾缂，也就是用深颜色的纬线勾出图案的轮廓，效果像工笔画一样。这里色调过渡的地方使用了长短戗。

小　龙：长短戗？这个名字也很奇怪。

龙教授：长短戗是在花纹由深至浅的变化中，利用缂丝线条的长短变化，使深色与浅色纬线交叉，得到一种自然晕色的效果。

看完缂丝画，小龙和大卫又被展厅中间的几把扇子吸引了。

大　　卫：这扇面上的牡丹太美了。

龙教授：这是一个仿制品，是仿朱克柔的《缂丝牡丹图》。牡丹象征着富贵吉祥，是丝织物中常用的纹样。

小　　龙：扇面上花瓣的颜色过渡得很自然，像是用画笔画上的一样。

龙教授：这是采用了缂丝技法中的木梳戗，也就是使用深浅颜色不同的纬线，从左向右或者从右向左织出木梳的纹路，呈现出花瓣渐变的效果。右边这个龙吐珠扇面上的云纹，也采用了木梳戗，用黑线和灰线层层缂织，呈现出由黑到灰的渐变效果。

小　　龙：龙吐珠象征着吉祥吗？

龙教授：是的，这种题材的缂丝扇面在元代很流行。

大　　卫：扇面是不是比较常见的缂丝作品？

龙教授：没错。缂丝扇既可以作为生活用品，也可以作为装饰品，无论古今都非常受欢迎。缂丝扇面上不仅会出现花鸟山水，还有书法文字。对于丝织品来说，书法作品的织作难度更高。

小　　龙：这里有好几个带书法的扇面呢。

大　　卫：这些缂丝的书法就像是写上去的一样，太了不起了。

Kesi Paintings and Fans

Xiaolong: David, look at this painting. It features soft colours and two cute, lifelike ducks.

Prof. Long: The painting is elegant and subtle. It's so typical of the Song Dynasty. This is *The Ducklings on the Lotus Pond,* a famous work by Zhu Kerou, a highly-skilled Kesi weaver of the Song Dynasty.

Xiaolong: How vivid the dragonflies, birds, and a pair of egrets are in the painting! The lotus flowers and leaves are also finely crafted!

David: It must have taken a lot of hard work to weave such an exquisite painting.

Prof. Long: You can say that again! It actually took eight years to finish this painting. It's said that more than two hundred kinds of silk threads were used. For weaving the Taihu stones alone, 25 kinds were used to present the different shades of blue.

Xiaolong: In that case, weavers themselves should also know about the basics of painting, right?

David: Or they could ask someone else to draw the draft before weaving it?

Prof. Long: For most of the time, it's the latter case, namely, copying the ready-made painting by weaving. But for Zhu Kerou, she was both a painter and weaver. Word has it that Zhu Kerou excelled at painting. Her Kesi works included a wide range of subjects, such as figures, flowers and birds. *The Ducklings on the Lotus Pond,* one of her seven surviving works, was housed in Shanghai Museum. It ranks as a first-class cultural relic and it's rare to see it on display.

Xiaolong: Dad, there're a lot of figures in this Kesi painting. I can't imagine how hard it was to weave them.

Prof. Long: Yes, it must have been a challenging process. This is a work from the Ming Dynasty. It's called *A Grand Birthday at Yaochi.* It describes the celebration of the birthday of Queen Mother of the West at Yaochi. All of the figures are immortals.

David: It's interesting that they all look different.

Prof. Long: Look at this one. It adopts some special Kesi techniques. For example, the technique of *gouke* (勾缂) is used at the edges. The weaver used dark-coloured weft threads to outline the pattern, which made the work look like a painting of fine brushwork. What's more, the technique of long-short *qiang* is used for colour transition.

Xiaolong: Long-short *qiang*? What a strange name!

Prof. Long: Well, the unique craftsmanship creates a natural sense of shading from darkness to lightness of patterns by interweaving long and short lines of dark and light weft.

Taking their eyes off the Kesi paintings, Xiaolong and David were attracted by some Kesi fans in the middle of the exhibition hall.

David: Look at this fan. How beautiful the peony is!

Prof. Long: Indeed. The fan was woven after Zhu Kerou's work *Peony*. The peony symbolises wealth and good fortune and is a common design element.

Xiaolong: The petals' colour is graded so naturally that it looks like the flower painted there.

Prof. Long: That's the technique of wooden comb-like *qiang*. It makes the patterns of dark and light lines look like the teeth of a wooden comb, through weaving dark threads from left to right and light threads from right to left or the other way round. Look at our right-hand side. It's also applied to the cloud pattern on the fan featuring a dragon spitting a pearl. The colour goes naturally from black to grey by the use of black and grey lines that step down in increments.

Xiaolong: Does the image also symbolise good luck?

Prof. Long: Exactly. Kesi works with this subject were very popular during the Yuan Dynasty.

David: Do most Kesi works take the form of fans?

Prof. Long: Yes. As a household item or a decorative piece of art, the Kesi fan has enjoyed enduring popularity. The patterns on fans include the artistic works of calligraphy as well as flowers, birds and landscape. Calligraphy, in particular, is not easy to be woven on silk fabrics.

Xiaolong: There're quite a few silk fans with calligraphy on them here.

David: Yes. And it looks as if the calligraphy was written on the fabric. That's amazing!

通经断纬

参观完缂丝织品，小龙、大卫和龙教授最后来到缂丝工具展厅。

小　　龙：爸爸，织造这些精美缂丝作品需要专门的机器吗？

龙教授：不用，最常用的平纹织机就可以，不过需要一些专门的缂丝工具，主要有梭子、拨子和毛笔。

小　　龙：我知道梭子是织纬线用的。拨子是干什么的？

龙教授：拨子形状像梳子一样，用来帮助把纬线打紧，盖住经线，不让它们露出来。

大　　卫：怎么还要用毛笔呢？

龙教授：用毛笔画样，就是先用毛笔在经线上勾勒出轮廓，然后用梭子按照画好的图案编织。

大　　卫：那主要工具还应该是梭子吧？

龙教授：是的，缂丝用的是一种小梭。织机上纵向排布的是经

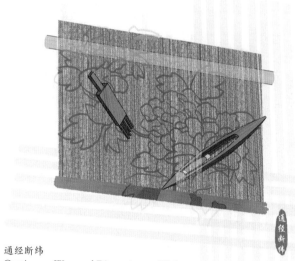

通经断纬
Continuous Warp and Discontinuous Weft

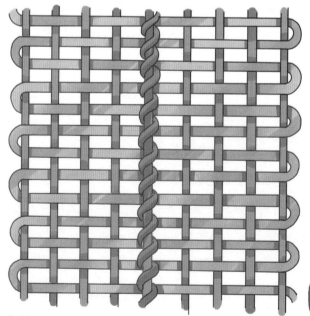

子母经　Mother-and-Child Warps

线，梭子上缠绕的是纬线。梭子带着纬线在经线间往复，织出花纹。

大　卫：这些梭子的样子很有趣，两头尖，底部平，像小船的样子。

龙教授：两头做得尖，是为了挑经，同时可以减少摩擦，方便来回穿梭；底部做得平滑，是为了防止穿梭时磨损下面的线。古时候还用象牙、牛角做梭子。光滑的材质可以减少对丝线的磨损。

小　龙：织出来的缂丝画那么漂亮，那缂丝使用梭子的方法一定也很特别吧？

龙教授：是的。一般织布是梭子穿过所有经线再返回，那叫"通梭"。但缂丝时，梭子常常是穿过一定数量的经线就返回，这叫"断纬"。"通经断纬"是缂丝特有的技法，就是经线贯穿整个布面，不同的梭子里装上所需的彩色纬线，按照不同的纹样图案，在不同的地方回返，分区分片缂织。

大　卫：经线和纬线用的丝线是一样的吗？

龙教授：不一样。缂丝的经线用生丝，纬线一般用染色的熟丝。生丝可以让布料挺括，色线负责缂出丰富的色彩。缂丝的纬线比经线粗。

小　　龙：如果频繁更换，那织出来的丝织品不是会出现很多接头，显得很杂乱吗？

龙教授：织工们有办法不让它出现杂乱的接头。缂丝的最后一道工序是修毛，就是把背面的线头一根一根剪断，使其与正面图案一样平整。背面的线头越多就证明图案的色彩越繁复，所以修毛时需要一根一根地贴根剪掉，既要有耐心，还得细心。前面的展品你们看到接头了吗？

大　　卫：好像没有看到。

龙教授：通经断纬织出来的花纹色彩正反两面一模一样，这是缂丝制品特色之一。

小　　龙：那怎么才能准确知道纬线该在什么地方换色？弄错了是不是就没办法修改了？

龙教授：这个不用担心。织造之前，织工会先布置好经线，然后将要织的花纹图案放在经线下面，用毛笔在经线上绘出图案的轮廓，然后才开始缂织。

小　　龙：听起来好像不难。

龙教授：但实际操作很不容易，要不怎么缂织一幅作品要好几年呢。

大　　卫：龙教授，除了通经断纬，缂丝还有别的技法吗？

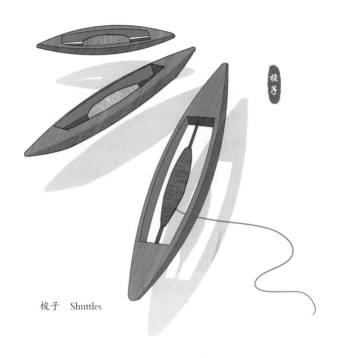

梭子　Shuttles

龙教授：如果只是通经断纬，织造中会出现纵向花纹裂缝。为了避免出现裂缝，需要运用两种重要技法——"子母经"和"搭梭"。

小　龙：为什么会出现裂缝？

龙教授：缂丝的时候，纬线依据纹样来回往复，并不贯穿整个织品，纬线遇到了不同花纹和色彩就要往回织，不同色彩的纬线不相交叉的话，就会产生裂缝。

大　　卫：那怎么解决这个问题呢？

龙教授：在织物上每隔一定距离，让两边的色纬互相搭接一下，就可避免形成裂缝，这就是搭梭。这个技法在缂织画作中使用得比较多。子母经是搭梭的进一步发展，主要用于汉字和印章图案的缂织。运用这种技法时，需要在主要经线，也就是母经上加拴一条细经线，这就是子经，然后再将不同颜色的纬线织在经线上。

小　　龙：这样的话，是不是要用到两个梭子？

龙教授：没错。具体的子母经技术还是很复杂的，需要结合实际操作过程才能看清楚、搞明白。

小　　龙：那这样做有什么优点呢？

龙教授：这个技术解决了缂丝中放大文字的问题。在这之前，缂丝作品中的文字都很小。这个技术出现以后，缂丝作品中的文字可以达到很大的尺寸。

大　　卫：工匠们的聪明智慧和钻研精神真让人佩服。

Continuous Warp and Discontinuous Weft

After visiting the Kesi fabrics, Xiaolong, David and Prof. Long went to their last stop. It was an exhibition hall of Kesi tools.

Xiaolong: Dad, are there any special looms for those exquisite Kesi works?

Prof. Long: No, the most commonly used tabby loom is good enough. But some Kesi tools are needed, like shuttles, plectrums and brushes.

Xiaolong: I know shuttles are used to weave weft threads. What about plectrums?

Prof. Long: These comb-shaped tools are used for tightening the weft threads so that they can cover the warp threads.

David: What about the brushes then?

Prof. Long: They're used for drafting. The outline of patterns should be first drawn on the warp threads. Afterwards, weavers use shuttles to weave the Kesi in line with the drawn draft.

David: So, shuttles are much more important in weaving?

Prof. Long: In a sense, yes. And by shuttles, here we mean small shuttles. The warp threads stretch vertically on looms while the yarns twined on shuttles are weft threads. With weft threads, shuttles are flitting back and forth through warp threads to weave patterns.

David: Lovely! A shuttle looks exactly like a tiny boat, with tapered ends and a flat bottom.

Prof. Long: Yes. The two ends are tapered to pass through the warp threads, reducing friction and facilitating the movement of shuttles. And the bottom is made flat and smooth to prevent threads below from fraying. In ancient times, some shuttles were even made of ivory or cow horns, because smooth materials can reduce the wear and tear on silk threads.

Xiaolong: I see. The Kesi paintings here are beautiful beyond description. Their way of using shuttles must be different from that of others, right?

Prof. Long: Exactly. In most cases, shuttles will run through all the warp threads before returning to the start. But when it comes to Kesi, the technique known as "continuous warp and discontinuous weft" is employed, and the shuttles will often go back after passing through a certain number of weft threads. The warp threads are across the loom and the shuttles with silk threads of various colours go back and forth in accordance with the specified patterns. As a result, the weft threads are woven in a discontinuous manner.

David: Are the weft and the warp threads actually the same kind of silk threads?

Prof. Long: Not the same. The warp is made of raw silk, which makes the fabric firm, while the weft is made of coloured and processed silk, which helps to create colourful patterns. And the weft is thicker than the warp.

Xiaolong: There'll be a lot of tangled ends left if the shuttles are frequently changed, right?

Prof. Long: Usually, the craftsmen have a way to fix it. The last step of Kesi is trimming. The thread ends on the back are scissored off one by one to make it as flat as the front. The number of thread ends on the back shows the complexity of the patterns. And they need to be cut off individually, which requires patience and care. We didn't see any ends in the previous works on exhibition, right?

David: No.

Prof. Long: One feature of the Kesi works is that both sides are the same in pattern and colour.

Xiaolong: How do the craftsmen know where to change the colour of silk threads? There is no going back when getting it wrong, right?

Prof. Long: No worries! Warp threads will be set in place first. Then with a brush, the weavers can paint the pattern on the warp following the draft under it. Only after that can the weaving process begin.

Xiaolong: It must be easier said than done.

Prof. Long: Indeed. Remember? A piece of Kesi work may take several years to complete.

David: Prof. Long, are there any other specific techniques for Kesi, in addition to the technique of "continuous warp and discontinuous weft"?

Prof. Long: Yes. During the weaving process, weavers need to use the techniques of mother-and-child warp threads and interlocked weft threads to avoid cracks in the longitudinal patterns.

Xiaolong: Cracks?

Prof. Long: As we know, the weft threads don't run through the loom. Instead, they are woven in and out according to the pattern. In other words, the weft of one colour will return as it encounters the weft of another colour. So, if different colour blocks don't cross, cracks will appear.

David: Are there any possible solutions?

Prof. Long: Absolutely. The first solution is to interlock two neighboring weft threads at regular intervals in the fabrics. This technique is called the interlocked weft, commonly used in Kesi paintings. The second one is the technique of mother-and-child warp. It's an extension of the first, mainly for weaving Chinese characters or seal patterns. With this weaving technique, one of the fixed warp threads is twined around by a flexible warp thread. And this fixed thread is the mother warp, while the flexible one is the child warp. Weft threads of varied colours are woven on them.

Xiaolong: If so, two shuttles are in need, right?

Prof. Long: Exactly. Actually, the technique is quite complicated. You may come to a better understanding of these technical terms in the practical experience of weaving.

Xiaolong: What are the advantages of using the technique of the mother-and-child warp?

Prof. Long: With this technique, characters on the Chinese silk tapestry can be of a larger size. Before that, the characters on Kesi works used to be quite small, and with this technique, the size of characters was no longer a problem.

David: The wisdom of ancient craftsmen and their commitments were really admirable.

结束语

中国是农业大国，创造出灿烂的农耕文明。桑蚕丝织技艺是其中极具代表性的一部分，很早就形成了从种桑养蚕到丝织产品的规模化生产。丝绸织造是古代中国人认识自然、利用自然的成果，其工艺渗透着中国人的聪明才智，产品及其应用表达了中国人的生活理念和审美情趣。丝绸是中国人民对人类文明的一项伟大贡献。

Summary

China has been a major agricultural country with a splendid agrarian civilisation in the past millenniums, in which sericulture and silk craftsmanship constitute a representative sample. China has achieved mass production from mulberry planting and silkworm rearing to silk weaving very early. Inspired by nature, the ancient Chinese invented the weaving of silk fabrics. And with their ingenuity and wisdom, exquisite textiles have come alive. All these products and their application reflect the life philosophy and aesthetic tastes of Chinese people. The silk fabric is Chinese people's great contribution to human civilisation.

中国历史纪年简表
A Brief Chronology of Chinese History

夏	Xia Dynasty			c. 2070—1600 B.C.
商	Shang Dynasty			1600—1046 B.C.
周	Zhou Dynasty	西周	Western Zhou Dynasty	1046—771 B.C.
		东周	Eastern Zhou Dynasty	770—256 B.C.
		春秋	Spring and Autumn Period	770—476 B.C.
		战国	Warring States Period	475—221 B.C.
秦	Qin Dynasty			221—206 B.C.
汉	Han Dynasty	西汉	Western Han Dynasty	206 B.C.—25
		东汉	Eastern Han Dynasty	25—220
三国	Three Kingdoms			220—280
西晋	Western Jin Dynasty			265—317
东晋	Eastern Jin Dynasty			317—420
南北朝	Northern and Southern Dynasties	南朝	Southern Dynasties	420—589
		北朝	Northern Dynasties	386—581
隋	Sui Dynasty			581—618
唐	Tang Dynasty			618—907
五代	Five Dynasties			907—960
宋	Song Dynasty			960—1279
辽	Liao Dynasty			907—1125
金	Jin Dynasty			1115—1234
元	Yuan Dynasty			1206—1368
明	Ming Dynasty			1368—1644
清	Qing Dynasty			1616—1911
中华民国	Republic of China			1912—1949
中华人民共和国	People's Republic of China			1949—

图书在版编目（CIP）数据

中国传统桑蚕丝织技艺．缂丝：汉英对照 / 郭启新，赵传银主编．-- 南京：南京大学出版社，2024.8
（中国世界级非遗文化悦读系列 / 魏向清，刘润泽主编．寻语识遗）
ISBN 978-7-305-26374-3

Ⅰ．①中… Ⅱ．①郭… ②赵… Ⅲ．①桑蚕丝绸－缂丝－丝织工艺－介绍－中国－汉、英 Ⅳ．① TS145.3

中国版本图书馆 CIP 数据核字（2022）第 234225 号

出版发行	南京大学出版社
社　　址	南京市汉口路 22 号　　邮　编　210093
丛 书 名	中国世界级非遗文化悦读系列·寻语识遗
丛书主编	魏向清　刘润泽
书　　名	**中国传统桑蚕丝织技艺．缂丝：汉英对照**
	ZHONGGUO CHUANTONG SANGCAN SIZHI JIYI. KESI: HANYING DUIZHAO
主　　编	郭启新　赵传银
责任编辑	张淑文　　编辑热线　（025）83592401
照　　排	南京新华丰制版有限公司
印　　刷	南京凯德印刷有限公司
开　　本	880mm×1230mm　1/32 开　印张 5.5　字数 114 千
版　　次	2024 年 8 月第 1 版　2024 年 8 月第 1 次印刷
ISBN 978-7-305-26374-3	
定　　价	69.00 元

网址：http://www.njupco.com
官方微博：http://weibo.com/njupco
官方微信号：njupress
销售咨询热线：（025）83594756

* 版权所有，侵权必究
* 凡购买南大版图书，如有印装质量问题，请与所购图书销售部门联系调换